Cannabis Crashes

Myths & Truths

BY

SCOTT MACDONALD

Distributed by Lulu: http://www.lulu.com/shop

Contact information of author:
Scott Macdonald
Scientist, Canadian Institute of Substance Use Research (CISUR) and Professor, School of Heath Information Science, University of Victoria
Email: scottmac@uvic.ca

Acknowledgments

This book is based on my 40-year career in research. I would like to thank Sonya Ishiguro, Katerina Perlova, Kylie Ransom, and my wife, Christine Glenn, who provided background research and reviews of drafts for this book. Content advice was provided by Mark Asbridge, Russ Callaghan, Brian Emerson, Bill Kerr, Tim Stockwell and Cameron Wild. Jeff Brubacher provided extensive content suggestions on earlier drafts. Justin Sorge helped produce figures for this book. Jeff Brubacher and Justin Sorge provided advice on the Pharmacokinetics chapter. Jonathan Woods designed the book cover. Amanda Farrell-Low and Judith Henstra provided editorial advice. I am indebted to the countless researchers in this area who have provided both formal and informal comments of various issues related to substance use and crashes.

Table of Contents

Preface

Over the past century, alcohol intoxication has been identified as the most important contributor to traffic fatalities and is strongly related to all types of automobile crashes (i.e. less harmful collisions). Thousands of studies showing consistent findings of ***performance deficits*** from alcohol form a strong evidence base for these conclusions. The advent of the quick and portable Breathalyzer test for alcohol ***impairment*** paved the way for effective interventions to reduce casualties from drinking and driving. Widespread deterrence-based legislation has demonstrated the effectiveness of prohibiting driving at various cut-off levels of ***Blood Alcohol Content (BAC)***, expressed as a percentage of alcohol in the blood. Scientists and most citizens accept these facts.

The story is considerably different with respect to cannabis. In this book, I will describe why knowledge about alcohol cannot directly be applied to cannabis. There is research on cannabis and performance deficits, as well as drug tests to detect cannabis use; however, many findings are dissimilar to those for alcohol, and the quality of the research is considerably worse. In many instances, authors from these scientific studies draw conclusions that are not justifiable from the methods used. Some of these faulty conclusions have become widespread beliefs in society, generating myths about cannabis use.

In Western societies of the 20th century, total prohibition of cannabis use and sales has been the predominant approach. More recently, nine US states (Alaska, California, Colorado, Maine, Massachusetts, Nevada, Oregon, Vermont and Washington) and Canada have legalized or plan to legalize recreational use of cannabis, and another 13 states have decriminalized its use. In preparation for legalization, the Government of Canada (2017) readied a backgrounder report that describes proposed legislative changes for laws in relation to cannabis and driving.

Canadian legislation permits roadside oral fluid (saliva) tests when law enforcement officers reasonably suspect drivers are under the influence of cannabis. Following a positive reading, officers could demand drug evaluation of a blood sample. A blood test reading of 2 to 5 ng/mL ***Tetrahydrocannabinol (THC)*** in ***whole blood*** would be subject to a summary (i.e. less serious) criminal conviction and a reading over 5 ng/mL could be subject to an indictable (i.e. more serious) offence. Those with both a BAC of over .05% alcohol and 2.5 ng/mL could be charged with an ***indictable offence***.

These legislative recommendations may be challenged under the Canadian Charter of Rights. The major question addressed in this book is **how to set cut-off thresholds for cannabis drug tests.** The first step in this process is to evaluate the ***validity*** of different cut-offs using an epidemiological approach. Societal values and legal guidelines for legislation can then be used to recommend legislation with an understanding of the validity of different drug tests. The aim of this book is to provide readers with objective scientific information about cannabis performance deficits and how to measure them.

A primary focus in this book is the validity of drug tests based on cut-off levels of ***compounds*** of biological fluids (i.e. blood, oral fluids and urine) in assessing performance deficits that could be considered impairment. In order to draw conclusions, the quality and cut-off values found in research related to alcohol risk assessment are compared to those associated with cannabis. I have summarized key issues that emerged from my research. For those who do not have a background in social science research, it may be worthwhile to review the ***glossary of terms*** to better understand some of the scientific concepts presented here. Any term the reader encounters in bold and italics can be found there.

Much of the material and my interpretation of studies presented in this book are derived from my career as a scientist and professor. I have published over 100 peer-reviewed papers, specializing in substance use and injuries, and have been an expert witness in over 20 court hearings involving drug testing in the workplace. During this process, the variety of papers I have read ranged greatly in quality.

I have witnessed the science behind our understanding of cannabis and traffic crashes evolve greatly over the past two decades. Initially, authors of a 1999 review concluded that: "There is no evidence that consumption of cannabis alone increases the risk of culpability for traffic crash fatalities or injuries for which hospitalization occurs, and may reduce those risks" (Bates & Blakely, 1999; p. 231). On the other hand, recent understanding drawn from a major meta-analysis concludes: "Acute cannabis consumption is associated with an increased risk of a motor vehicle crash..." (Asbridge et al., 2012; p. 1). These disparate interpretations reflect the progression of our knowledge; both interpretations were well founded, given the quality of the scientific studies at the time. However, many other conclusions have been drawn since 1999 that, in my opinion, are not justified by the scientific methods used. In this book, I identify some of the issues that have created myths in the study of cannabis and vehicle crashes.

This book is intended for several audiences, including students, policy makers, lawyers and experts in the field of substance use and crashes. I review in detail some methodological and statistical issues important for understanding conclusions from studies and the degree to which they are justifiable. As well, I review relevant studies on cannabis and crashes and the validity of biological drug tests to identify those who are impaired.

CHAPTER 1

Introduction

CANNABIS LEGISLATION

In 2018, Canada became the second country in the world to legalize cannabis, after Uruguay legalized it in 2013. Some U.S. states, including Colorado, California and Washington, have also permitted the sale of cannabis, although federal regulations still make cannabis illegal in the United States. Legalization is a response to widespread acceptance of the drug and normalization of its use. Years earlier, the Canadian Senate Special Committee on Illegal Drugs Laws indicated that laws are a source for normative rules that should be used sparingly while respecting the freedoms of individuals to seek their own well-being (Nolin and Kenny, 2002). As well, they noted that criminalization is costly to enforce and largely ineffective in deterring use. Cannabis was criminalized in Canada in 1923, galvanized by a movement spearheaded by Emily Murphy, a police magistrate, who described cannabis users as raving maniacs, liable to kill or indulge on any form of violence to others (Hathaway and Erickson, 2003, p. 467). Impartial research on cannabis has been limited until recently and policy makers have largely lacked evidence-based information to make sound decisions (Committee on the Health Effects of Marijuana, 2017). As a consequence, many myths and unknowns about cannabis existed under criminalization, which impede sound policy.

Other less drastic options than legalization are available, such as decriminalization, where possession is not subject to criminal convictions but those caught could still be subject to penalties, such as fines. Many countries have decriminalized cannabis, including Portugal, Switzerland, Italy, Mexico, Russia and Norway. Decriminalization is sometimes considered a half-way measure because a substance's use is not condoned, but tolerated.

The Canadian government's decision to legalize has been met with opposing viewpoints. Those against legalization have argued it will provide greater access to minors by normalizing pot and say that more road deaths will occur due to cannabis-impaired drivers. Those for legalization see it as an end to an overly invasive intrusion of a human's right to choose their own destiny.

Legalization involves a political trade-off. A tougher stance on some cannabis harms, such as impaired driving, may appease detractors and not overly alienate voters who staunchly supported legalization. Opinions of voters may have more influence on legislation than empirical evidence of minimizing harms. Yet science can be useful to provide a good basis for rationalization of different types of laws. The Canadian government has passed legislation to address cannabis-impaired driving, described in more detail in Chapter 2. The laws proposed include measures to detect impaired driving at low blood THC thresholds similar to many other countries worldwide. In this book, I focus on the empirical studies on the relationship between cannabis and ***impairment***[1] with the aim to better understand what different laws aimed at minimizing cannabis-impaired driving are likely to achieve.

TRAFFIC CRASHES DUE TO ALCOHOL AND CANNABIS

Impairment by alcohol while driving is a major contributor to traffic crashes. Recent figures from the U.S. show alcohol-impaired fatalities accounted for 31% of all traffic fatalities in 2014. In Canada, 22% (about 400) of fatally injured drivers were impaired above a .08% BAC in 2012, and 33% had a positive reading for alcohol use (International Transport Forum, 2016). The number and proportion of alcohol-related crashes have been steadily declining over the past few decades in Canada (Brown et al., 2013), but appears to have levelled off since 1998 in the U.S. (Fell and Voas, 2014). A roadside survey of drivers in a British Columbia, Canada, found 7.8% had been drinking, with 2.7% above .08% BAC, much lower than the proportion of drivers in fatal crashes (Beirness and Beasley, 2010). Such data, along with several high-quality ***case-control*** studies, point to alcohol as a major risk factor for traffic fatalities and collisions.

The quality of data for cannabis and traffic crashes/fatalities is significantly weaker than for alcohol. Researchers agree that driving while intoxicated by cannabis increases the risk of car accidents; however, the magnitude of this risk varies, depending on the geographic location of the study, type of crash (i.e. fatality, injury, or property damage),

---

[1] See the section "PERFORMANCE, IMPAIRMENT AND SAFETY" in this Chapter for a more detailed description of impairment.

biological cut-off level, and data collection methods. By all accounts, the average risk is low compared with alcohol. A recent study estimated that cannabis-attributable crashes have caused about 75 deaths in Canada (Wettlaufer et al., 2017), much lower than the number of traffic fatalities caused by alcohol. Again, the quality of the data for cannabis-related crashes is poor, and this study likely overestimated the number of deaths by using a low THC cut-off of 5 ng/mL in oral fluids, which the authors indicate corresponds to about .2 ng/mL THC in blood. Drivers near this cut-off would be unlikely to have caused crashes due to cannabis.

In the U.S., for the period between 1993 and 2010, 29.3% of drivers in fatal crashes were tested for drugs (Wilson et al., 2014). About 11.4% of drivers tested positive for any drug, and cannabis-positive cases accounted for about one third of those cases (Wilson et al., 2014). In Canada, the percentage of drivers involved in a fatal collision tested for drugs is higher at 77.4%, with about 40% testing positive for a substance other than alcohol, but tests specifically for cannabis were not reported (International Transport Forum, 2016). Another Canadian study of blood tests among fatally injured drivers over a 10-year period found 13.9% tested positive for drug use (Beirness et al., 2013). None of these studies have a comparison group of drivers not involved in crashes and therefore risks cannot be estimated.

A limitation of positive blood tests for cannabis is that impairment has not been determined. Universal testing with disaggregated data for each drug and detailed concentration levels is often unavailable. Our understanding of the relationship between drug concentration levels in the blood and impairment or accident risk remains poor.

Aside from major differences in the quality of the research, the issue of defining impairment is central to both cannabis and alcohol. In Canada and the U.S., impairment from alcohol has a legal definition, typically defined as BAC cut-off of .05% or .08% alcohol with legislation that prohibits driving at these levels. Studies show that practically all people at these levels have reduced skills needed for driving (Fell and Voas, 2014). Experimental ***laboratory studies*** have found performance deficits at lower BAC levels, but the extent to which drivers at these lower levels could be causally implicated in crashes is largely unproven. In this book, impairment refers to legal thresholds of alcohol concentrations of .05% and .08%. Performance deficits are defined more broadly to include any decrements in psychomotor, perceptual, or cognitive functioning. In this book, I aim to find equivalent impairment thresholds for cannabis as those for alcohol in terms of traffic-safety risks.

The ***prevalence of use*** of alcohol and other drugs can be estimated from surveys, (Canadian Alcohol and Drug Use Monitoring Surveys (CADUMS) from 2004 to 2011; Health Canada, 2015). These data show slight declining trends in use for cannabis and alcohol in

Canada from 2004 to 2011. In 2011, the population with past-year use was estimated at 78% for alcohol and 9.1% for cannabis. Prevalence of past-year cannabis use appears to be increasing moderately with 12% reporting any use in 2013, while alcohol use was largely unchanged from 2013, at 77% prevalence in 2015.

A question of importance is how frequently does the population engage in driving while under the influence of alcohol or cannabis? About 1.9% of surveyed adults in Ontario, Canada reported they drove within one hour after smoking cannabis (Walsh & Mann, 1999). By contrast, Beirness and Davis (2007) reported 11.6 % of Canadians admitted to driving within an hour of having two or more drinks. More recently, Statistics Canada reported that 14% of cannabis users with a driver's licensed said they drove within two hours of consumption (Statistics Canada, 2018).

This question also has been addressed by random roadside studies of drivers. Beirness and Beasley (2010) conducted a random roadside study where 78% of drivers stopped consented to a Breathalyzer test for alcohol and an oral fluid test for cannabis. About 8.1% had a positive alcohol test and 4.6% had a positive cannabis test. Given that oral fluid tests for cannabis have a much longer detection window (possibly days) than the alcohol Breathalyzer test, the differences between driving under the influence of cannabis or alcohol are likely greater than reflected by the percentages in this study. Compared to cannabis, alcohol contributes a much greater burden in terms of number of injuries and deaths due to crashes. Some believe that the reason for this fact is that most people drink and only a minority use cannabis. Although more people drink than use cannabis, as shown in this book, the absolute harms from alcohol are greater than for cannabis.

Both alcohol and cannabis are similar in terms of governmental interventions to control their use. Historically, alcohol was prohibited in Canada between 1856 and 1919; the provinces repealed prohibition beginning in 1924, with Prince Edward Island the last to abolish in 1948 (Hallowell, 1988). Today, several Canadian communities, primarily indigenous reserves, forbid alcohol use and sales. In the USA, prohibition existed from 1919 to 1933. During periods of prohibition, there was a polarization of attitudes toward alcohol until the scales finally tipped in favour of repeal (Tyrell, 1997). This polarization appears to be related to the emergence of various myths about alcohol in terms of its benefits and harms. One myth that emerged during prohibition was about alcohol's medicinal benefits. During that period, alcohol could be obtained legally with a doctor's prescription. Prescriptions were possible for a wide range of ailments, such as cancer, indigestion and depression (Okrent, 2010). Today, prescriptions like these for alcohol are rare. In terms of harms, prohibitionists also promoted some outrageous claims. For example, claims were levelled that alcohol was distilled from noxious ingredients including hemlock, opium and excrement, or that drinkers could spontaneously combust (Kelly, 2014). Following prohibition, polarized opinions still exist; however, alcohol regulations have become less stringent as alcohol is socially accepted.

Today, cannabis in Canada and several U.S. states is in a period of transition from prohibition to legalization, replete with polarized views and murky claims. On the anti-prohibition side, we have seen the emergence of medical cannabis for a wide variety of ailments, similar to what occurred during prohibition of alcohol. Although some medical uses have been scientifically proven, others have not (Watson et al., 2013). Opposing these views, prohibitionists have linked cannabis use with immigration, poor school performance, and claims that users would become maniacs liable for violence (Broughton, 2014). Legislation aims to balance these opposing societal views.

The Canadian government has recommended ***per se*** laws against driving under the influence of cannabis that are not dissimilar to many countries. In this book, I review these laws by drawing comparisons to laws against drinking and driving. I look at performance deficits with the use of alcohol and cannabis, and assess the utility of different interventions for detecting their use among drivers and workers. I also review the different methods of detecting alcohol and drug use through biological tests. A major focus of this review is the validity of the different tests to detect impairment given evidence of performance deficits. I examine the research evidence for detecting alcohol impairment and compare the approaches and findings to cannabis. As will be seen, the differences are substantial. At the end of the book, I speculate on the reasons and repercussions for these different standards.

RESEARCH QUESTIONS

In this book, I address three major questions:

1. What is the relationship between cannabis use and performance deficits?
2. What is the validity of biological tests for alcohol and cannabis to impairment?
3. How effective are alcohol and cannabis per se laws at improving traffic safety?

There are several pertinent issues to address regarding the relationship between cannabis and performance. First, it is important to understand the nature of potential risks associated with alcohol and cannabis use. This includes the acute effects of cannabis, as well as potential performance deficits such as withdrawal effects, hangover effects, and long-term cognitive deficits. Second, I will examine how to assess these deficits through biological drug tests. A final critical issue is the assessment of test validity, along with using test results for providing objective means for defining impairment.

Epidemiological principles for conducting research related to these topics are included. These principles have been developed and articulated to provide a guide on how to conduct research to ensure objective conclusions are drawn. Unfortunately, many research studies have major limitations, often unacknowledged by the authors. In good research, limitations of studies should be noted by authors so that readers can properly assess the meaning of research findings. Nevertheless, major limitations are not always highlighted by researchers, either because the authors are unaware of the limitations, or they are aware but wish to present their findings in a positive manner. I will identify the flaws in some studies and demonstrate why the conclusions by authors are not justified by the methods used.

Disciplinary focus

In this book, I recommend research from an epidemiological and biostatistical perspective approach to assessing the validity of diagnostic tests for cannabis impairment, and studies on cannabis, performance and crashes. Epidemiology and biostatistics is a discipline aimed at understanding the causes of diseases and intervention approaches. A major concern with epidemiology is the minimization of ***biases*** in research designs that can distort findings. Biostatistics involves objective analyses of data. The discipline of epidemiology and biostatistics is ideally suited for the study of the relationship between road crashes and cannabis. In this context, epidemiology is suitable for assessing the validity of biological tests to detect impairment or performance deficits (Gordis, 2014). Also, epidemiology methods prescribe optimal procedures for unbiased research designs, including laboratory studies and ***observational studies*** (Gordis, 2014). Finally, biostatistics aims to provide an objective analysis of data collected in studies. These approaches are both descriptive (a balance of measures of ***central tendency*** and ***dispersion***) and analytic to compare relationships among one or more variables. I focus on issues that are particularly relevant to the issue of the relationship between crashes and cannabis use, and the measurement with biological tests of impairment associated with increased crash risk. Since research on substance use and crashes is multi-disciplinary, some limitations of studies may originate from research disciplines that do not emphasize the issues of importance in epidemiology and biostatistics.

PERFORMANCE, IMPAIRMENT AND SAFETY

The focus of this book is performance deficits related to alcohol or cannabis use, and how these deficits increase the risk of traffic crashes. I draw a distinction between impairment and performance deficits.

- ***Performance deficits*** refers to the degree to which deficits occur under different conditions, which may or may not translate into dangerous driving.
- ***Impairment*** refers to a specific threshold at which driving a vehicle due to performance deficits is dangerous for nearly all people. For alcohol, this threshold is a BAC of .05% or greater.

Impairment implies some level of performance deficits that practically all people would experience; these deficits would also significantly increase crash risk. This definition is similar to that for alcohol impairment, described by Fell and Voas (2014). One reason for drawing this distinction is because impairment is typically a legal definition which refers to a specific alcohol concentration that constitutes a legal offence. Performance deficits may also occur at lower levels of consumption, but are not at a sufficiently great magnitude to meaningfully increase crash risk, or are not subject to penalties. Reference to any performance deficit as impairment obscures the difference between the two. Impairment relates to the choice of thresholds or cut-offs where drivers could be subject to penalties. This distinction has not always been made by researchers. For example, Moskowitz and Florentino (2000) define any performance deficit as impairment. In this book, a distinction is made between performance deficits and impairment. The purpose of defining impairment is to seek a meaningful threshold to classify individuals as either impaired or not impaired. On the other hand, performance deficits can be treated as a ***continuous variable,*** with deficits ranging from mild to severe.

An important distinction of deficits is the difference between an absolute standard versus a relative one. A person's skill and performance in operating a vehicle varies considerably from one person to the next, and performance for any individual can vary based on a number of factors at different times. Some people are more competent drivers than others. It stands to reason that when drivers experience different types of performance decrements, their ability to drive will be affected in different ways. Excellent drivers may become only good drivers, whereas average drivers may become poor when experiencing deficits. As well, a great deal of variation exists in terms of how drivers cope with these deficits in that some may successfully compensate whereas others may not.

What this means in real terms is that some drivers may still be capable of driving while experiencing various risk factors for crashes. Absolute standards refer to definite criteria needed to drive a car, such as understanding the rules of the road and being able to manoeuvre a vehicle satisfactorily. While it is likely preferable to implement an absolute standard rather than a relative one, this is impossible in practical terms for impaired driving.

Numerous risk factors have been noted to increase crash risk, such as aggressiveness, impulsivity, disrespect for authorities and depression (Macdonald, 1989). Stressful life events, such as the death of a family member and marital separation, are related to increased likelihood of serious traffic crashes (Lehman et al., 1987; Lagarde et al., 2004). Fatigue is also known as a risk factor for traffic crashes (Dawson and Reid, 1997). Although risk factors can be identified with empirical research, the identification of deterrence-style interventions to prohibit drivers with these conditions have not been feasible. Only educational approaches are feasible. These aforementioned issues pose challenges for defining substance impairment, and suggest that a fair biological threshold should be valid enough to correctly classify individuals into meaningful groups.

CONCLUSION

In this chapter, parallels and differences are noted between alcohol and cannabis in terms of traffic-safety risks and potential intervention approaches. In order to assess the magnitude of these safety risks and ways of measuring risk, an epidemiological approach is recommended. A distinction is drawn between performance deficits, which are continuous in nature, and impairment, a legal term that refers to a cut-off that can distinguish groups.

CHAPTER 2

Alcohol and cannabis legislation and impairment assessment

TYPES OF ALCOHOL AND CANNABIS DRIVING LEGISLATION

Impairment also has a legal definition, which varies considerably from jurisdiction to jurisdiction. These legal definitions are reviewed in this chapter. Worldwide, there are three major legal approaches for addressing alcohol or cannabis impaired driving: (1) per se cut-offs, (2) behavioural methods and (3) zero tolerance.

(1) **Per se laws** establish cut-off thresholds of substance concentrations in biological specimens (i.e. either blood, urine, oral fluids or breath), and state that these are sufficient evidence that an individual is guilty of driving while impaired. No additional information on performance by individuals is needed for conviction. If drivers refuse to provide a biological sample demanded by police, they can be charged for impaired driving.

Per se laws are the most common approach worldwide for both alcohol and cannabis. These laws exist in North America and the European Union for alcohol at .05% and .08% BAC. These cut-offs are consistent with research evidence that practically all people at this level show substantial performance deficits, and epidemiological studies indicate drivers at this level become a traffic-safety risk (Fell and Voas, 2014).

For cannabis, THC per se cut-offs are common in many countries and seven U.S. states (Wong et al., 2014). Most jurisdictions prescribe cut-offs based on whole blood although some prescribe blood ***serum***, urine or oral fluid tests. Concentrations cut-offs of THC are most common. Some jurisdictions have laws prohibiting the ***metabolite*** of THC, ***Carboxy-THC (THCCOOH).*** The strengths and limitations of these different biological specimens and detection of different compounds are discussed more fully in subsequent chapters. Typical per se laws forbid driving at thresholds between 2–5 ng/mL THC in whole blood or serum.

(2) **Behavioural criteria** laws to assess alcohol impairment based on performance measures rather than biological tests. Considerable efforts to operationally define these criteria have been made in North America with the development of the ***Standardized Field Sobriety Test (SFST)***, which can be used as evidence for Driving While Impaired by alcohol offences in Canada (i.e. a breathalyzer is not needed). The sobriety test includes three basic tasks: the horizontal gaze nystagmus, walk and turn heel to toe along a straight line, and the one-leg stand. Failure on a certain number of components of these tests will result in being charged. In North America and most other countries, behavioural criteria can be used to convict drivers of impaired driving in addition to the per se laws.

Similarly, for cannabis and several other drugs, the ***Drug Recognition Expert (DRE)*** program, also called the *Drug Evaluation and Classification Program* (DEC) was implemented in Canada and many U.S. states (Beirness et al., 2007; Porath-Waller, et al., 2009). The process involves a series of physical and psychomotor tests, and concludes with toxicological testing of a sample of a blood, urine or oral fluid (Royal Canadian Mounted Police, undated). This standardized procedure is used to identify both individuals under the influence of drugs and the type of drug causing the observable impairment. Thus, the DRE process is a sequential testing approach (i.e. behavioural symptoms plus a drug test) as opposed to the behavioural criteria used for assessing alcohol impairment, which by itself is sufficient to be charged by police.

A recent review paper of cannabis and driving legislation worldwide found that nine European countries had laws that include greater penalties for behavioural symptoms of impairment and lesser penalties for particular per se cut-offs, typically from THC concentrations in whole blood (Wong et al., 2014). Specific legislation exists in 16 European countries; 14 of these countries have legal THC limits of 2 ng/mL or below from whole blood or serum. In addition, three US states reported legal THC limits in blood of 2 ng/mL or below. Canadian laws of the 2 ng/mL cut-off are therefore roughly in line with several other countries. Whether or not these per se laws are consistent with existing scientific evidence is the focus of the next chapters.

As mentioned, this system is not perfect for cannabis detection as mistakes occur. A question of importance is whether the new proposed laws will be an improvement. How accurately will the laws be for identifying drivers who are a safety hazard? Will the laws make the roads safer? Will they be enforceable at both the police and judicial level?

(3) **Zero-tolerance** laws prescribe penalties at a threshold level of zero or the low levels of detection at which alcohol or drug can accurately be detected and can be thought

of as per se laws that are not linked to performance criteria. For alcohol, a BAC of .02% is considered zero tolerance because research evidence does not show that a BAC level at .02% constitutes impairment, whereas for cannabis, 1 ng/mL THC in blood is considered zero tolerance. Zero-tolerance laws are not premised on the presumption of impairment or any other subjective judgment about a driver's behaviour. Countries with cut-offs of .02% alcohol or below are considered zero tolerance. A few Eastern European countries, the Czech Republic, Hungary and Slovakia have absolute zero limits for alcohol. Other countries, such as Poland, Finland and Sweden, have limits of .02%, which allows for insignificant amounts of alcohol.

Six zero-tolerance countries in Europe and one U.S. state use THC limits of 1 ng/mL in whole blood or ***blood serum*** (Wong et al., 2014). Sanctions for cannabis users who drive is expected to reduce the likelihood of impaired cannabis driving and make the roads safer.

THE HISTORY OF ALCOHOL AND CANNABIS LEGISLATION

Alcohol

Over a hundred years of laboratory evidence shows that psychomotor abilities deteriorate considerably with the ingestion of alcohol. Increases in BAC levels are significantly related to decreases in performance (Jellinek & McFarland, 1940; Moskowitz, 1973; McKim, 1986; Coambs & McAndrews, 1994; AMA Council on Scientific Affairs, 1986; Moskowitz & Florentino, 2000). The question of importance was how to use this research to make the roads safer. Most legal interventions established in the early 1900s made impaired driving an offence based on behavioural grounds of impairment. In the U.S., the first laws against drinking and driving were implemented in 1910 in New York. In 1939, Indiana introduced legislation that allowed evidence of BAC levels as corroborating evidence of alcohol impairment but had little value on its own. In Canada, the first criminal code legislation against drinking and driving was instituted in 1921. Initially, drinking and driving was penalized as a less serious summary offence; it was not until many years later that it became an indictable (more serious) offence. In 1912, the first drinking-and-driving laws were implemented in Norway and several Scandinavian countries followed over the next two decades (Ross, 1975).

It wasn't until 1936 that the first per se law was implemented in Norway and Sweden, using an early version of the Breathalyzer. Although Robert F. Borkenstein often is credited with inventing the Breathalyzer that assesses BAC, the first devices using breath tests to detect BAC date back to the 1870s (Wutke, 2014). The early science focused on developing an accurate breath sample that mirrored alcohol content in the blood. Studies showed that Breathalyzer readings corresponded very closely to performance deficits. The Borkenstein device, coined the Breathalyzer, was different from early prototypes in that it was portable, making it ideal for roadside use.

In 1964, at the same time as the Breathalyzer emerged, Borkenstein and his team conducted the first large scale, case-controlled epidemiological study that established a ***relative risk function*** demonstrating an exponential relationship, starting at .08% alcohol, between BAC levels and crash risk (Borkenstein et al., 1964). This study, along with the Breathalyzer, paved the way for the first per se laws in North America, and made driving at a given BAC without behavioural grounds sufficient evidence for criminal penalties (Gore, 2010). These interventions greatly reduced alcohol-related collisions by increasing the certainty of apprehension.

In the late 1960s, per se laws were instituted in U.S. states and Canada for driving with a BAC of .08% alcohol or more, based on a Breathalyzer reading. In early years, only tests conducted at a police station were considered evidentiary. Today, police are required to have reasonable grounds based on behavioural observations to demand a driver's breath sample (or a blood sample for those unable to provide a breath sample). Roadside Breathalyzer readings are considered sufficient evidence. Over time, the penalties for drinking and driving became more severe and additional refinements, such as inclusion of penalties for refusing to take a Breathalyzer, were introduced. The Canadian government has recently tabled legislation to allow police to order a breath test without probable cause (Government of Canada, 2017). Drinking-and-driving cases are the largest single-offence group (they made up about 12% of all criminal charges in 2008), with about 53,000 drinking-and-driving cases heard every year in Canada (Kenkel, 2015).

Internationally, virtually every industrialized country has a per se law. The first per se law was implemented in Norway in 1936, which make it an offence to drive with a BAC of .05% (Voas and Lacey, 1990). Per se legislative cut-offs are highly variable in different countries, ranging from 0% to .15%, with most countries applying limits of .05% to .08%. Lower limits in Scandinavian countries are likely related to the strength of ***temperance movements*** in these areas. In Sweden, a BAC of .02% is punishable as a criminal offence. Higher limits are noted in some U.S. states where alcohol use is widely accepted as a societal norm. Per se laws are confusing for some, as the amount of alcohol that different people can drink while remaining under BAC thresholds of .05% or .08% vary and depends on the type of beverage consumed. This ***variability*** is based on the amount of ***ethanol*** consumed, body weight, metabolism, and time since last drink. An advantage of low per se laws (i.e.

under .05% alcohol) is the simplicity of avoiding any drinking before driving. A drawback is that drivers at low BAC levels may be punished even though they do not necessarily represent a meaningful safety risk.

Laurence Ross, a prolific researcher who studied the impact of numerous per se legislations worldwide, noted positive short-term effects with their introduction (Ross, 1984). A more recent review of evaluation studies on countries that introduced or lowered legal BAC limits, confirms there are beneficial traffic-safety effects, at least in the short-term (Mann et al., 2001). Such research has helped galvanize social movements, such as Mothers Against Drunk Driving (MADD), which called for stronger deterrence-style legislation and increased funding to address this issue (Ross, 1984). Legislation then, and most worldwide legislation since, is based on the theory of *deterrence*. ***Deterrence theory*** posits that the certainty, swiftness and severity of punishments for offenders will change unwanted behaviours (Ross, 1982). Ross found that perceived certainty of punishment was more effective than severity of punishment. Consistent with this conclusion, Norway and Sweden abandoned mandatory jail sentences for drinking and driving above certain limits (Ross and Klette, 1995).

In Canada and the U.S., there are limits to both the certainty and swiftness of punishment that can be achieved. Enshrined within the Criminal Code laws are the rights to due process, protection from unreasonable search and the rights to defend oneself (Kenkel, 2015). The tenet that "beyond a reasonable doubt" is needed to convict someone of a criminal offence indicates that criminal conviction of an innocent person is worse than not convicting someone who is guilty. Furthermore, while criminal code laws mostly incorporate the principle of ***mens rea***, meaning that individuals must have criminal intent, this principle has been largely bypassed with per se laws. These factors impede the introduction of even tougher laws and are barriers to deterrence, as the judicial process is time-consuming. Substantial proportions of individuals charged under per se laws are never convicted. More leeway exists within provincial legislations, which mandate traffic safety and are guided by principles of public health more than individual rights. Most provinces have penalties for driving at a BAC of .05%, with penalties less severe than those under the Criminal Code of Canada but fewer safeguards for due process. Total abstention from alcohol is required for young/novice drivers and ***alcohol ignition locks*** are common for previously convicted drivers. British Columbia, for example, introduced a separate set of laws for drinking drivers known as the Immediate Roadside Prohibition Program, laws that are not feasible at a federal level. This provincial policy essentially bypassed the Criminal Code of Canada, as police could immediately suspend driver's licences for three days among drivers with BAC levels between .05% and .08% alcohol (Macdonald et al., 2013). The policy was found to be effective in reducing three types of alcohol-related collisions over a two-year period.

Since per se limits vary so widely among countries, from zero to .15% alcohol, a question of importance is what cut-off is optimal. A low cut-off of .02% or less sends a clear-cut message that no alcohol in the blood is acceptable while driving (Klette, 1983). Although the Swedish will staunchly defend their laws and much of the public support it, their effectiveness in preventing alcohol-related crashes compared to higher limits of .05–.08% alcohol has not been clearly established (Ross, 1975; Klette, 1983). However, the preponderance of the research evidence does indicate lowering per se cut-offs, typically from the .08–.10% range to .05% alcohol, produces beneficial safety effects that may be temporary.

Cannabis

As was the case with alcohol a hundred years ago, many logistical issues related to the development of effective interventions against impaired driving by cannabis are being grappled with today. In Canada and the U.S., it is illegal to drive a vehicle while impaired by cannabis; however, to determine when drivers are actually impaired by cannabis is much more difficult than for alcohol. In Canada and most U.S. states, convictions are based on behavioural grounds of drug impairment (discussed in more depth in this chapter), which are difficult to prove.

Legalization of cannabis in Canada brings new penalties for drivers with THC concentration in whole blood. These are:

- 2 to 5 ng/mL → summary conviction/offence
- Over 5 ng/mL → indictable offence
- BAC over .05% alcohol and over 2.5 ng/mL → indictable offence

Legislation in Canada will permit roadside oral fluid tests when law enforcement officers reasonably suspect that a driver has ingested cannabis (Government of Canada, 2017). This policy may be expanded to include suspected use of other drugs as well. A positive saliva sample would allow officers to demand a blood sample to conduct an evaluation for THC concentration.

Above, I have reviewed legislative approaches for alcohol or cannabis-related driving. Such legislation can be assessed on empirical grounds, if the purpose is to determine limits where drivers can be considered a traffic-safety risk. This can be achieved from laboratory studies that assess the validity of biological tests, observational studies in the real world on the relationship between cut-off levels and crash risk, and evaluation studies on the implementation of laws using cut-offs.

THE PURPOSE OF BIOLOGICAL TESTS

Biological tests to assess impairment offer several advantages over tests using behavioural methods. They are more objective and easier to administer, and ideally, they can provide a single objective measure that is closely related to performance deficits at a magnitude known to be associated with increased crash risk.

Assessing the validity of biological tests is based foremost on the purpose for which the test is intended. Drug tests have been proposed for two different purposes: (1) to identify drug users, or those who have used cannabis and other drugs within specified time periods, and (2) to identify those who are impaired by drugs. The first purpose is concerned with detection of a drug concentration indicating drug use. The second purpose is the assessment of a particular drug concentration threshold to classify individuals as impaired, and is the focus of this book.

A valid test is one that correctly identifies people who are impaired (i.e. have performance deficits that increase safety risk) from those who are not. In order to accomplish this goal, we must have a good idea of how to define impairment. For alcohol, this has been achieved through countless laboratory and observational field studies. The laboratory studies show that virtually all people have measurable and meaningful performance deficits at .05% BAC (Fell & Voas, 2014). Also, field studies show the risk of crashes increases rapidly at BAC levels of .05%. Similar criteria for assessing the validity of blood THC can be applied to cannabis. After defining the meaning of impairment as an ***operational definition***, this ***gold standard*** can be used to assess the validity of a drug test in laboratory studies. The gold standard is an external source of truth regarding the magnitude of deficits known to be associated with increased crash risk.

Validity of a drug test can be measured in several dimensions, as outlined by Gordis (2014) and illustrated in **Table 1**. ***Accuracy*** is defined by the sum of the correctly identified subjects with respect to impairment (i.e. true positives and true negatives) divided by the total sample. Two common measures are ***sensitivity*** and ***specificity***. Sensitivity is the ability of the test to identify correctly those who are impaired, whereas specificity refers to the ability of the test to identify correctly those who are *not* impaired. Two directly related indicators, ***false negatives*** and ***false positives***, are defined by the respective formulas of (1 – sensitivity) and (1 – specificity). Another important indicator is ***positive predictive value***, or the probability that someone who tests positive is truly impaired. The ***Kappa*** statistic is often used to assess the degree of agreement between a test and a gold standard. The ideal situation is a perfect test where all of the persons identified as impaired are truly impaired, and all those identified as non-impaired are truly not impaired. Regrettably, this ideal is unrealistic. A Kappa value of over .75 is optimal (excellent agreement). Scores between .4 and .75 are intermediate to good agreement and under .4 is poor agreement (Gordis, 2014; p. 110).

Table 1: *Different indicators of validity of a biological test for impairment*

Test result at a particular cut-off	Impaired	Not impaired
Positive	True positive (TP)	False positive (FP)
Negative	False negative (FN)	True negative (TN)

Source: Adapted from Gordis (2014, p. 91)

Accuracy = (TP+TN)/ (TP+FP+FN+TN)
Sensitivity = TP/ (TP+FN)
Specificity = TN/ (TN+FP)
Positive Predictive value = TP/ (TP+FP)
Negative Predictive value = TN/(FN+TN)
Kappa = (Percent observed agreement – Percent agreement expected by chance)/ (100% – Percent agreement expected by chance).

Drug concentrations are detectable at various levels as a ***continuous variable*** and cut-offs are chosen to reflect a reasonable trade-off. Assuming that the drug concentration level bears some relationship to impairment, even a weak one, lower thresholds will increase false positives (reduce specificity) and decrease false negatives (increase sensitivity). Conversely, higher thresholds will decrease false positives and increase false negatives. If the consequences of false negatives are very great, then a lower threshold is recommended, and if the consequences of false positives are great, then a higher threshold is warranted. The threshold chosen depends on the degree of importance placed on the consequences of the two types of errors. For alcohol and drug testing, this trade-off includes values of safety versus individual freedoms, the validity of the tests, and the effectiveness of the different interventions. As recommended by Wald and Bestwick (2014), a useful approach is to specify the acceptable sensitivity rate for a given specificity rate (or false positive rate). This issue is largely unaddressed in research. Typically, overall accuracy is acceptable if the Kappa value exceeds .75 (Gordis, 2014).

Ideally, assessment of the validity of a biological test for impairment should be made against a gold standard. This gold standard would typically be performance deficits associated with crashes. Unfortunately, as described in Chapter 5, performance deficits are measured in such a wide variety of ways that the precise definition is still largely undefined for cannabis deficits. Research on this topic is superior for alcohol compared to cannabis. Nonetheless, research has been conducted on assessing the validity of the field sobriety test battery against BAC levels and the DRE against positive or negative THC, reviewed below.

Alcohol

Interestingly, few laboratory-style studies have been conducted using the epidemiological approach to assess the validity of the Breathalyzer to detect impairment. Some of the early research was directed to develop more accurate psychophysical tests that best discriminated at a cut-off level of .10% alcohol (Burns and Moskowitz, 1977). This research recommended the one-legged stand, walk and turn and alcohol gaze nystagmus as the three most valid combined indicators of alcohol impairment. These are direct measures of performance deficits caused by alcohol and can be thought of as the gold standard for impairment. The vast majority of people who are not inebriated would find these tests simple to pass.

Stuster and Burns (1998) assessed the validity of the Standardized Field Sobriety Tests (SFST) for alcohol impairment against a BAC threshold of .08%. Seven police officers trained in the behavioural approach rated 297 drivers on the road, according to a scoring system that estimated BAC levels. I re-arranged the data from the Stuster and Burns (1998) study to be consistent with my epidemiological approach (see **Table 2**). I have assessed the validity of a BAC cut-off of .08% alcohol against expert behavioural symptoms. Overall, a BAC cut-off of .08% alcohol is valid compared to SFST. Positive predictive value is excellent at 98%, meaning a driver with a BAC of over .08% has a 98% likelihood of being judged impaired by the SFST. The Kappa value is .75, indicating excellent agreement. Specificity, at 94%, is greater than 90% for sensitivity, which corresponds to a larger percentage of false negatives than false positives. Findings from this study provide evidence of excellent validity of .08% alcohol for impairment.

Stuster and Burns (1998) also compared officers' assessments with BAC cut-offs of below .04% alcohol with between .04 and .08% alcohol and calculated accuracy at 80%. Based on this data, I calculated Kappa at .36, indicating poor agreement. Unfortunately, all the cases above .08% alcohol and others are missing from the table, and I was not able to replicate the table for a simple .04% cut-off. However, from the information available, the level of agreement between officers' assessments of a .04% cut-off with BAC cut-offs declines considerably into the unacceptable range.

Table 2: *The validity of .08% BAC cut-off compared to Standardized Field Sobriety Tests*

	SFST assessment for a 0.8% cut-off		
.08% cut-off	**Impaired**	**Not impaired**	**Total**
Positive	210	4	214
Negative	24	59	83
Total	234	63	297

Source: Adapted from Stuster and Burns (1998).

Accuracy = 91%
Sensitivity = 90%
Specificity = 94%
Kappa= .75

Positive Predictive value = 98%
Negative Predictive value = 94%

Burns and Anderson (1995) compared police decisions to arrest in Colorado where it is illegal to drive with a BAC over .05% with known BAC cut-offs of .05%. Results from this study are presented in **Table 3** with an analysis of the validity. As can be seen from the table, the Kappa value drops to the moderate range at .61 when compared to Stuster and Burns (1998). Specificity declined fairly substantially. Sensitivity increased, consistent with expectations with a lower threshold. My overall assessment is that the test has a moderate degree of validity. Based on these findings, criminal code penalties are not warranted; however, lesser penalties under provincial legislation or summary convictions without criminal records based on a public health approach are justifiable.

In another study, Stuster (2006) compared police observations based on SFST (the gold standard) with BAC cut-offs at .08% alcohol. They calculated Kappa values for each of seven police officers with values that ranged from .61 to 1.0, indicating intermediate to excellent agreement. Although more studies of this type are warranted, the findings of this study along with the large body of additional laboratory studies comparing BAC levels with performance indicators (reviewed in Chapter 5), do indicate current BAC cut-off levels of .05% and .08% to determine impairment are justified in North America.

Interestingly, all the aforementioned studies indicated their purpose was to validate the SFST, and not a particular BAC cut-off. In other words, BAC was treated as the gold standard of impairment rather than SFST. This indicates how entrenched BACs have become a valid indicator of impairment. BAC is a biological measure that is used for convenience and expediency and as such should be validated against external gold standards of impairment. The SFST represents the best behavioural measure of alcohol impairment and preceded wide use of BACs. In any case, the strong agreement between SFST and BAC indicates validity of both.

Table 3: *The validity of .05% BAC cut-off compared to Standardized Field Sobriety Tests*

	Behavioural assessment of .05% cut-off		
.05% cut-off	**Impaired**	**Not impaired**	**Total**
Positive	163	21	184
Negative	12	38	50
Total	175	59	234

Source: Adapted from Burns and Anderson (1995)

Accuracy = 86%
Sensitivity = 93%
Specificity = 64%
Kappa= .61

Positive Predictive value = 89%
Negative Predictive value =76 %

Cannabis

The Drug Recognition Expert (DRE) program, also called the Drug Evaluation and Classification Program (DEC) program, uses a 12-step procedure to assess cannabis impairment, which represents our best measure to date; however, it is inferior to the SFST for alcohol. Biological drug tests have been used to assess the validity of the DRE approach. This approach is much less valid for cannabis impairment than for alcohol. The DRE process includes 12 steps involving clinical measures, such as blood pressure, pulse and pupil size which are not focused on behavioural impairment and have not been properly validated. Although some clinical signs, such as elevated blood pressure and pulse, may be more common among those with cannabis impairment, these conditions are common among those without impairment and appear to be poor measures for cannabis impairment. Other steps are more directly focused on performance, such as the finger-to-nose and modified Romberg balance test, indicators of divided attention. These divided-attention tests appear to be fairer criteria to classify individuals as road worthy. Research is needed to assess whether drivers who do not pass these tests have the necessary skills for safe driving. This type of abbreviated approach is one that could be considered as an intervention approach by the Provinces. If a gold standard of cannabis impairment is to be achieved, it will be through the development of indicators of direct performance measures, and ideally, signs that are specific to cannabis impairment. The final step is a drug test with blood, oral fluids or urine, which is also a poor biomarker of performance deficits that could be considered impairment, as demonstrated in Chapter 3 of this book.

In a review of the validity of the DRE program based on both laboratory and field studies, Beirness et al. (2007) identified nine studies that implemented an epidemiological approach for the validity of DRE officers' assessment for identifying drivers under the influence of drugs. However, all the studies were limited in that they compared DRE assessments with drug tests that are unproven to show impairment. In four of these studies, subjects were administered cannabis and DRE assessments were made of behavioural indicators of impairment. We do know that, for most people, the acute effects of cannabis produce deficits that make driving a car unsafe (as shown in other sections of this book). In laboratory studies, subjects were administered both cannabis and ***placebos*** under highly controlled experimental conditions and subjects were tested shortly thereafter, when impairment is likely (although some studies used very low doses). Although Kappa values were not reported by Beirness et al. (2007), overall accuracy, sensitivity and specificity are reported to assess validity. As shown in **Table 4,** the accuracy of identifying those who were administered cannabis is better than chance, but poor overall, ranging in accuracy from 39.7% to 63.6%. These studies do not inspire confidence that the DEC approach is valid for accurately detecting impairment, given the controlled conditions of the laboratory studies.

In field studies in real-world conditions, DRE assessments were compared to toxicological findings that were either positive or negative, based on urine, oral fluid or blood tests. These studies are flawed because we don't know if these biological measures had any valid relationship with impairment, a limitation identified by Beirness et al. (2007). Based on what we know about these drug tests, the studies determined whether the DRE assessment were valid to detect use, not impairment. Noting that accuracy rates were better for field studies than laboratory studies, Beirness et al. (2007) speculated that police officers had the benefits of gaining additional information from an initial investigation and that the dose may be greater in real-life field settings compared to laboratory doses. Although these explanations are reasonable, the DRE assessment in all the field studies were compared against biological tests, which detect prior use (not necessarily currently under the influence), whereas in all the laboratory studies, cannabis was administered shortly before assessments when being under the influence is likely. Higher accuracy rates in field studies could also be related to the detection of users rather than those impaired, whereas the laboratory studies were more specifically aimed at detecting those under the influence.

Table 4: *Accuracy, sensitivity and specificity of DRE assessment with biological tests*

Author	Accuracy (%)	Sensitivity (%)	Specificity (%)	Type of test	Lab or field study
Bigelow et al., 1985	63.6	48.8	92.7	Drug administered	Lab
Heishman et al., 1996	56.0	53.1	61.1	Drug administered	Lab
Heishman et al., 1998	39.7	30.4	59.1	Drug administered	Lab
Shinar & Schechtman, 2005	41.7	49	69	Drug administered	Lab
Compton, 1986	74.6	59.7	86.4	Blood	Field
Preusser et al., 1992	75.4	78.4	73.2	Toxicology reports	Field
Hardin et al., 1993	90.1	93.8	82.6	Urine	Field
Smith et al., 2002	79.9	80.5	76.6	Toxicology reports	Field

Source: Adapted from Beirness et al. (2007)

Regardless of these aforementioned limitations, only one study (Hardin et al., 1993) produced indicators of validity (i.e. accuracy, sensitivity and specificity) that were even close to the study by Stuster and Burns (1998) for alcohol impairment. The DRE protocol is not a scientifically valid technique for proving that an individual has consumed a particular class of illicit drug resulting in impairment (Smith et al., 2002), but officers are able to identify drug classes with an accuracy better than chance (Beirness et al., 2007). Nonetheless, an advantage of the DEC approach is increased specificity and fewer false positives, (Gordis, 2009) because ***sequential tests*** are used rather than compared a single measure. Sequential tests are more appropriate for diagnostic than screening purposes. The DEC process is cumbersome for charging suspected drug-impaired violators, and another drawback is a low conviction rate for drug-related charges (Solomon and Chamberlain, undated).

In a similar study by Beirness et al. (2009) actual DRE evaluations and toxicology reports were compared. The authors concluded the DREs in Canada are accurate in that DRE substances matched toxicological reports cannabis use in 87.3% of the cases. Similar to limitations of the 2007 study, we cannot say whether the subjects were unsafe to drive – only that they met the criteria of impairment under Canadian laws. The authors note that subjects are asked about recent drug use for investigations, which can have an influence on decisions. Also, the authors do not know whether they received a comprehensive list of all DRE cases. Other studies have assessed the validity of various aspects of cannabis impairment but none were found that used the epidemiological methods that I propose.

In a study by Papafoutiou et al. (2005), the authors assessed a modified SFST that included an additional performance test of head movements compared against time points after cannabis administration at five, 55 and 105 minutes. Behavioural assessments of impairment were best at five minutes after use in the high-dose condition (56.4% classified as impaired) and declined after 105 minutes (38.5% classified as impaired). This study only focused on sensitivity (not specificity or overall validity) and direct comparisons were not made with blood tests, so the study is of marginal value for understanding the validity of blood tests. No studies were found where behavioural symptoms of impairment were compared directly with THC blood tests cut-offs intended to assess impairment.

Hartman et al. (2016) compared 302 DRE cases of those with confirmed impairment based on DRE symptoms with 302 non-using controls on a number of measures. These measures included the modified Romberg balance test, walk and turn, finger to nose and one-leg stand. These tests aim directly at performance deficits related to cannabis impairment and normally non-impaired people should find them easy to pass. Sobriety tests can have greater validity when the test relates directly to one's ability to drive a car. In this respect, the common sobriety test of the one-legged stand and walking a straight line toe to toe appears to be an excellent measure of driving capability. These tasks are exceedingly simple to complete for most sober people. However, when intoxicated, they become difficult. Harman et al. (2016) also tested several physiological symptoms, such as pulse, systolic blood pressure and pupil size. Several variables significantly distinguished the two groups; however, such measures do not appear to have good validity as they are very indirect measures of use, and do not necessarily indicate performance deficits. Although this study is useful for identifying variables that can be used to identify cannabis impairment, no differences in measures were found among cases with blood THC above 5 ng/mL and those below, a finding that doesn't support the hypothesis that THC cut-offs in blood are a valid indicator of impairment. Another study reported similar findings where Field Sobriety Tests were useful in detecting cannabis impairment but blood tests for THC were not (Declues et al., 2016). This study suggests a single THC per se limit of 5 ng/mL in blood will not have good validity.

CONCLUSION

Virtually every industrialized country has per se laws with BAC cut-offs ranging from .02% to .15%. In Canada and most U.S. states, it is a criminal offence to drive with a blood alcohol concentration (BAC) of .08%. Under federal criminal law, citizens have rights that restrict the ability for law enforcers to prosecute offences, such as the right to due process and reasonable search. In Canada, provincial governments have additional leeway to implement more stringent drinking-and-driving laws. Highway and Traffic Acts in many provinces contain laws specifying lower BAC limits, commonly .05% alcohol, and immediate consequences for those who break the law. These laws have largely been successful in reducing alcohol-related crashes.

The laws worldwide for driving under the influence of cannabis appear to parallel those for alcohol and crashes. For impaired driving by both alcohol and cannabis, biological samples are used to convict individuals. Research shows that a BAC threshold of .08% has excellent validity against objective behavioural symptoms, such as the SFST, and a BAC cut-off of .05% has moderate validity. As BAC levels decline, the validity of SFST become much less reliable to discriminate those who have consumed alcohol from those who have not. For cannabis, the 12-step DRE approach has good accuracy against biological tests in field studies, and poorer validity against known drug administration (Beirness et al., 2007). If DRE assessments are considered the gold standard for impairment, they have not been validated against a reasonable threshold for impairment using any type of biological test. Most of the 12 steps do not identify behavioural symptoms of cannabis impairment, and there are too many steps, making it fairly impractical as an intervention approach.

Also, the modified Romberg balance test, walk and turn, finger to nose and one-leg stand have some promise as valid tools for police officers in the field to detect impairment. Symptoms of cannabis impairment are very subtle and difficult to detect. Since symptoms of being under the influence of cannabis are so difficult to detect, driving under the influence of cannabis may not be as hazardous and driving under the influence of alcohol. As demonstrated in this book, much of what we know about alcohol-impaired driving and detection cannot be applied to cannabis-impaired driving. I compare our scientific knowledge of alcohol with cannabis on issues related to performance deficits and how to detect them throughout these pages.

For per se laws to be based on thresholds for cannabis impairment, they should be validated with experimental studies. There are major gaps in the literature as no studies were found where behavioural symptoms of impairment were compared with a threshold above the limit of detection to assess the validity of any THC cut-off using a standard epidemiological approach.

CHAPTER 3

Detection approaches for alcohol and cannabis

ANALYTIC METHODS

Alcohol

The main method of detecting BAC levels is with portable Breathalyzers. These devices are typically based on an infra-red and/or a fuel cell device (Wigmore, 2014). Breathalyzers are considered to give highly accurate approximations of BAC levels, as outlined below, although they are not perfect. For example, to be valid, subjects should not have anything to drink at least 10 minutes before the test and must be able to provide a full breath sample.

Cannabis

Several analytic methods exist for the detection of prior cannabis use, but two major approaches deserve special attention: *screening* and *confirmatory* tests. Screening tests, based on an ***immunoassay*** procedure, use antibodies to detect the presence of drugs and their metabolites (Dyer & Wilkinson, 2008). These immunoassays generally lack accuracy to detect a specific type of compound. Rather, they detect several types of compounds that have similar molecular structures, but the combination and concentration levels of each compound are unknown. A major limitation of these screening tests is reduced accuracy. Screening tests are used due to their lower costs and results that can be quickly obtained and interpreted on-site. Recent Canadian legislation for driving under the influence of cannabis includes oral fluid screening tests, but the analytical methods are not specified.

A key challenge with respect to field testing novel methods is the validity, expense, time and portability of different analytic tools. The immunoassay approach can provide rapid results within minutes and without a laboratory. Additionally, the immunoassay approach is relatively inexpensive and highly portable. A major limitation is the of lack validity, as described above. The confirmatory approaches with ***chromatography and spectrometry*** methods generally have excellent validity but require more expensive non-portable laboratory equipment and time-consuming analysis.

Confirmatory tests are near perfect for detection and identification of specific drug compounds and use chromatography and spectrometry methods (Jaffee et al., 2008). These methods are often used to compensate for the inaccuracy of immunoassays approaches. To analyse urine, gas chromatography and mass spectrometry ***(GC/MS)*** are typically used, whereas oral fluids are analyzed using liquid chromatography and tandem mass spectrometry ***(LC/MS/MS)***, methods that are common and considered equally valid. As mentioned, these confirmatory approaches effectively identify specific compounds but require more expensive non-portable laboratory equipment and analysis are time-consuming. The validity of these tests relates to detection of concentration levels of the compound or metabolite being targeted, not behavioural symptoms, such as being under the influence of a drug. Validity regarding how concentration levels relate to impairment is discussed in later chapters.

SUBSTANCES DETECTED

In the material below, I focus my review on the elimination rate of various drugs based on drug concentration levels. These estimates are derived from ***controlled dosing studies*** where subjects are administered drugs and concentration levels of different chemicals in the drug are measured at different time points.

Alcohol

The Breathalyzer is designed specifically to detect ethanol, the single molecule of alcohol, C_2H_5OH.

Cannabis

Drug tests for cannabis can be used to detect the presence of various compounds and metabolites. Cannabis is extremely complex, with over 400 different compounds, many with psychoactive properties (Adams & Martin, 1996; Atakan, 2012). The main psychoactive component of cannabis is delta-9-tetrahydrocannabinol (THC), which has been most studied and the most promising compound to choose as a cut-off in relation to impairment. ***Carboxy-THC*** (THCCOOH), a THC metabolite, is typically detected with urinalysis and used to identify drug users rather than those impaired. Like THC, carboxy-THC can be stored in fat cells for days and possibly weeks in heavy daily users. ***Hydroxy-THC (11-OH-THC)*** is another major metabolite of THC. Compounds and metabolite concentrations are typically expressed in nano-grams per millilitre (ng/mL). The focus of this book will be mainly on the detection of THC and these metabolites. There are other psychoactive components in cannabis, identified by Sharma et al. (2012), but little research has been conducted on their impact on performance. Another component in cannabis, ***cannabidiol (CBD)*** has been widely researched for its medicinal effects but is not psychoactive, and cannot be considered a viable biomarker for impairment.

DETECTION PERIODS FOR DIFFERENT BIOLOGICAL SAMPLES

Controlled dosing studies, where subjects are administered substances and concentration levels are measured at different time intervals after use, provide useful data on the drug concentration elimination rate and detection periods over time. Knowledge of elimination rates can provide only a rough approximation of concentration levels associated with typical impairment periods. In the following chapter, I compare the typical impairment periods for alcohol and cannabis with the concentration levels. In subsequent chapters, I review studies of paired detection tests and performance measures.

Various biological samples can be analysed for the presence of cannabinoid compounds: blood, breath, oral fluids, urine, sweat and hair. Sweat testing and hair testing are not reviewed in detail here. Briefly, hair testing is suitable for assessing exposure to drugs up to 90 days. Sweat testing is typically used for monitoring drug use (through a transdermal patch) over extended periods of time (Cone et al., 1994).

Differences exist among biological sample types in relation to what can be learned from the test results, especially when detecting performance deficits and impairment. Detection periods for specific compounds vary enormously based on the type of biological sample. For most compounds, the detection periods are the shortest in blood, and longest in hair. Oral fluids have shorter detection periods than urine. Since performance deficits from the acute effects of drugs are short-lived, blood tests are frequently cited as the best matrix to assess impairment for all substances. It is important to note that a specific concentration level in one biological sample is not equivalent to that in another type of biological sample.

The concentration level of compounds and metabolites between different samples vary considerably for individuals at the same point in time. For example, oral fluids generally contain 20 to 60 times higher concentrations of THC soon after use, compared with blood samples from the same person.

In this section, I describe the detection periods for alcohol and cannabis in blood, breath, urine and oral fluids, with a focus on elimination rates in the different biological samples and the validity of alternative samples to blood.

UNDERSTANDING STATISTICS

The importance of measures of dispersion—In reviewing the controlled dosing studies, various types of summary statistics are presented by authors. Summary statistics are useful to provide simpler understanding of the sample distributions. Measures of ***central tendency*** (mean, median and mode) and measures of ***dispersion*** (***standard deviation, coefficient of variation (CV),*** and range) are common descriptors. For reviewing controlled dosing studies, measures of dispersion have more relevance when entertaining the possibility of using drug tests to discriminate those impaired from unimpaired. Measures of central tendency tend to obscure differences between individuals. Since with drug tests, we draw conclusions about individuals, it is important to know how well group averages can be used to draw inferences of individuals—and to do this, we need to know how similar people are based on measures of dispersion.

In controlled dosing studies, all subjects are administered the same amount of cannabis and therefore should have similar THC concentrations. Large differences between individuals or within the same individual over time are sources of error that indicate the likely impracticality of drug tests, unless the variations are closely related to impairment. In the following material, these terms are used and I provide an interpretation of their meaning.

When interpreting the degree of dispersion in a sample, I typically examine three primary measures: standard deviation, coefficient of variation and range of the sample. If the standard deviation is extremely small in relation to the mean (i.e. under 5% the size of the mean) or extremely large in relation to the mean (i.e. greater than the mean), the degree of variability of the sample is either extremely low or extremely high, respectively. Essentially this is a ballpark estimate of the CV. Coefficient of variation (also called the relative standard deviation) is a universal measure that provides a measure of dispersion for any sample with different measures (i.e. units of analysis). The CV is calculated by dividing the standard deviation of the sample by the mean. **Table 5** provides a ballpark interpretation of the meaning of CVs in terms of dispersion.

Table 5: *Interpretation of coefficients of variation (CV)*

Coefficient of variation (CV)	Degree of variability
Under 5%	Extremely low
Between 5–15%	Low
Between 15–85%	Average
Between 85–95%	High
Over 95%	Extremely high

Note: CV = standard deviation/mean

Correlations—Another statistical test introduced in this chapter is the linear ***correlation***, which is assessed with a ***correlation coefficient (r)*** and ***coefficient of determination***. In linear correlation, a straight line of best fit is plotted based on minimizing the distance between all the data points. Other types of relationships, such as curves, between points can also be calculated but are less commonly used. ***Statistical significance***, determined by probability (p) values, provides an objective measure of the likelihood that a relationship is caused by something other than random chance. The p-value is related to both the sample size (i.e. number of subjects) of a study and the variability between measures. Virtually any relationship, no matter how weak, can be significant with a large enough sample; however, it does not mean that the significant relationships are necessarily meaningful. Tests of statistical significance are necessary to demonstrate a relationship, but such tests are not sufficient to say the relationship is good, moderate or poor.

Correlation coefficients (r), are used for comparing two continuous variables and range from -1 to 1 with a plus or minus sign, which represents the direction of the relationship. A ***correlation of determination*** (r^2) provides an objective means to define the strength of relationships between two continuous variables (i.e. ***effect size***). If the purpose of the correlation is to demonstrate that one measure is valid against another measure, such as between oral fluid and blood THC concentrations (or between blood alcohol content and Breathalyzer tests for alcohol), r^2 can be interpreted in the same way as Kappa coefficients if the purpose of the test is to substitute one type of measure for another (i.e. .75 is excellent, between .4 and .75 is intermediate and under .4 is poor). The r^2 value has more intrinsic meaning than a correlation coefficient (r). When r is squared, it can be interpreted as the percent of variation in one variable that is shared with a second variable. If the purpose is to show that a strong relationship exists between two variables, such as

THC blood levels and performance deficits, then less stringent criteria are needed than for substituting one type of test for another. As a general guideline, an r^2 value of .5 or higher is desirable. Some studies of the relationship between BAC levels and performance deficits exceed this threshold (see Dawson and Reid, 1997).

When assessing whether a statistical relationship is meaningful, it is important to take note of both the statistical significance and the effect size. The probability value is the likelihood that a given strength of a relationship is due to chance, and for this value to be significant, it should be below 5%, or $p<.05$. The likelihood of ***probability values*** being significant is based on both the sample size and the variability in the measures. If statistical significance is achieved, then it is useful to examine the effect size of a relationship to see if it is meaningful, given the purpose of the test.

An assumption of correlations is that all observations are independent of one another (i.e. assumption of ***independent measures***). In some studies, multiple paired measures of the same individuals are taken and used to calculate a correlation coefficient. This procedure is incorrect and will typically produce inflated correlation coefficients, giving a distorted impression that the relationship is stronger than reality. This inflation occurs because the variability among pairs of observations of the same individuals is more consistent than those between different individuals. If researchers have done this, it is best to disregard the paper because the findings cannot be interpreted properly.

TYPES OF BIOLOGICAL SAMPLES

Alcohol

Blood—Blood tests are considered the best biological sample for assessing blood alcohol content or BAC (Lindberg et al., 2007); however, even with blood tests, readings can be misleading in certain situations. Misleading results can be obtained from diabetics, and in those taking certain cough medicines or certain herbal supplements.

Blood is analysed as three major mediums: (1) ***whole blood***, (2) ***serum*** or (3) ***plasma.*** Serum and plasma are obtained by centrifuging a whole-blood sample into its constituent parts. Serum and plasma are more common in medical facilities, whereas whole blood is typically mandated for legal purposes. In drinking-and-driving cases, blood samples of injured drivers can be taken for medical reasons and can later be subjected to subpoena by police, and it is important to understand the difference. The ratio between alcohol in serum/plasma and whole blood is fairly consistent. Rainey (1993) reported 12 empirical studies on this issue, with a mean ranging from 1.10 to 1.21. Rainey (1993) calculated that 99% of the population would have a ratio of .90 to 1.49. The low standard deviation and low range in ***conversion factors*** indicates moderate variations between individuals.

The elimination rate of ethanol for an average individual is about .016% absolute alcohol per hour, or between one half and one standard Canadian drink per hour (Andreasson et al., 1995). The rate of alcohol elimination from the blood for a given individual is very constant.

Breath—Breath tests were the first test developed for detection of alcohol and are advantageous given they are more convenient and less intrusive than drawing blood. In order to assess their validity, studies have been conducted that compare blood and Breathalyzer tests for subjects who have consumed various amounts of ethanol. In a laboratory study conducted with 799 subjects, the average blood/breath ratio was 2407 (95% ***Confidence Interval (CI)***: 1981–2833) (Jones & Andersson, 1996), indicating low dispersion of the ratios. More recently, Jaffe et al. (2013) conducted a study that addressed issues of potential measurement bias in prior studies. The study included 61 subjects with paired blood/breath alcohol content measures. Treating blood as the gold standard with a cut-off of .05% BAC, breath tests had high sensitivity (97%) and specificity (93%). The blood/breath ratio of alcohol on average is about 1 to 2400, but is not a constant factor both between and within individuals over time (Jones, 2010). In order to compensate for errors in conversion and the fact that the ratio increases over time, Canadian laws have adopted a conservative ratio, which means BAC calculated based on breath alcohol results is lower than the actual BAC measured in blood

Extensive research has been conducted that demonstrates portable breath test findings are closely correlated with blood tests. Alcohol breath readings have been validated against blood concentration through statistical calculation of correlations between the two from paired readings of different subjects. An example of findings from one study of suspected impaired drivers by alcohol is provided (Zuba, 2008). The coefficient of determination (r^2) in this study was .88. This value means that breath concentrations accounted for between 88% of the variation in blood readings. This represents a strong relationship and validates the breathalyzer as an acceptable alternative test for a blood *test* (see Chapter 7, Strength of association, for interpreting r-values). A similar study by Gainsford et al. (2006) of four different breathalyzers found nearly identical relationships. Findings similar to these have been found in several other studies. The breathalyzer, while not perfectly accurate, is an excellent substitute for blood tests.

Oral fluids—Oral fluids can also be tested for alcohol and are reported to have a near perfect correlation (r=.98) with blood tests (Haeckel and Peiffer, 1992). The feasibility of using oral fluid tests was established over 20 years ago through ratios for oral fluid and blood concentrations that demonstrated extremely low variability between individuals (Coefficients of variation, CV, between 3.2 and 4%) (Jones, 1995). Oral fluid tests offer a good alternative for individuals who are unable to provide a breath sample, as these tests are not highly invasive. Alcohol detection with oral fluid tests is gaining interest, as they can be used to detect alcohol and several other drugs simultaneously.

Urine—Urine tests to detect alcohol are much less accurate than breath and blood tests for BACs. One limitation of urine tests is that alcohol may not be detectable for up to two hours after drinking. This is especially problematic for assessing impairment. Another issue is unacceptable variations between individuals for the urine/blood conversion, which was studied several decades ago (see Payne et al., 1967 for a review). The British Medical Association recommended a conversion factor of 1.32/1 once the peak concentration level had been passed. The range between individuals' urine/blood ratio has been found to be from .92 to 2.32, which indicates poor agreement.

In conclusion, although alcohol can be detected easily in urine two hours after drinking, accurate estimation of BAC levels cannot be made. Considering the invasiveness of urine collection (compared with breath tests) and its poor accuracy, urine tests are rarely used. Other biological samples such as sweat and eye vapours have been studied as possible specimens for impairment, but are not widely used.

Cannabis

Cannabis can be detected through analysis of various biological specimens, including urine, oral fluid, sweat, breath and blood. There is strong agreement among experts that of these biological samples, blood tests are the most valid for detection of concentration levels associated with impairment for alcohol and cannabis (Wolff et al., 1999; Lindberg et al., 2007). However, drawing blood requires specialized training, is invasive and the blood needs to be analysed by laboratories. As well, as demonstrated in this paper, blood needs to be taken almost immediately when impairment is suspected. If the sample can't be analysed soon after collection, sodium fluoride can be used to preserve the sample (Kapur, 2009). The detection time (i.e. the time from last use to the test) for all drugs will vary among individuals, regardless of the testing technology, depending on the dose taken, frequency of use, route of administration, duration of use, tolerance, metabolic rate and cut-off levels of the tests (Verstraete, 2004; Dyer and Wilkinson, 2008).

Blood—As with alcohol, blood is considered the best (although clearly not optimal) biological specimen for correlating concentration levels of compound to performance under cannabis conditions, due to its shorter detection periods compared to oral fluid or urine. Research shows that concentration levels of THC from the same individual differ greatly depending on the medium of blood being analysed: either whole blood or serum/plasma. Concentration levels are thought to be roughly equivalent for serum and plasma. Both of these specimens contain higher drug concentrations than whole blood. Furthermore, empirical evidence suggests that whole blood/plasma ratios vary considerably between individuals. For example, Schwilke et al. (2011) found significant differences (p=.003) in the whole blood/plasma THC ratios between subjects. Desrosiers et al. (2014) reported a median whole blood/plasma conversion factor of .68 with a range between individuals from .31 to 1.1, showing a poor relationship. Although the preponderance of research clearly shows whole blood to serum or plasma ratios are less than one, the large range in ratios also indicates poor ***reliability*** in the conversion factors between individuals. Plasma or serum levels should not be used to draw inferences of whole-blood levels for THC for an individual. This poor conversion factor is a major limitation of numerous empirical studies, as serum/plasma cannot accurately be generalized to whole blood and vice versa for any individual. This issue is a major limitation if said studies are used to draw conclusions for a cut-off in blood to identify impaired individuals.

Despite the substantial degree of variability between individuals and over time for the same individuals, conversion factors from serum/plasma to whole blood are frequently cited. The most commonly cited conversion factor is .5 (1 ng/mL THC in whole blood is equivalent to 2 ng/mL in serum) reported by Grotenhermen et al. (2007). Couper and Logan (2004) report the average conversion of whole blood to serum/plasma for THC is .55 (or 1 ng/mL in whole blood is equivalent to 1.82 in serum or plasma). No empirical studies to support either of these conclusions were cited. For drawing a rough ballpark estimate of the average conversion for a **group**, a conversion rate of .5 is acceptable; however, when it comes to an **individual**, this conversion is unacceptable for estimating whole-blood THC concentration levels from serum/plasma or vice versa because the variability of the ratio between people is so great.

In terms of elimination times after smoking cannabis, a controlled dosing study by Huestis et al. (1992) recorded blood plasma cannabinoid concentrations of THC, carboxy-THC and hydroxy-THC at 34 different time points for eight subjects with high-dose (3.55% THC) and low-dose (1.75% THC) marijuana cigarettes. Peak concentration levels of THC were recorded under 15 minutes for every subject. Peak concentration levels of hydroxy-THC occurred at a longer time period (under 30 minutes), and peak concentration for carboxy-THC was longer than both at four hours or less. Detection periods, the time periods from last use which each compound can be detected, are shortest for THC and longest for carboxy-THC. Since THC is the main psychoactive ingredient of cannabis and the acute effects of cannabis are short-lived, THC is the most studied compound as a possible biomarker for impairment.

In another study by Karschner et al. (2009), long-term heavy smokers of cannabis were monitored for seven days of abstinence in an in-patient setting. One daily user had a THC blood concentration level of 3.7 ng/mL six days after abstinence. A later study by Desrosiers et al. (2014), concluded that 16.7% of frequent cannabis users had THC concentrations exceeding 5 ng/mL at 30 hours after use.

Hunault et al. (2014) administered three different doses of cannabis to participants. For each dose, THC blood levels increased rapidly within the first 15 to 30 minutes after smoking. In the high-potency condition (69.4 mg THC), THC concentration levels spiked at about 210 ng/mL in serum blood within the first half hour, and in the low potency condition THC spiked at about 140 ng/mL. Then the THC levels declined rapidly by 90% within the first hour, followed by a very gradual decline afterwards. By four hours, THC levels declined to low levels (although exact numbers were not provided in the report) and remained at low levels for the eight hours of the study.

Other studies of THC (1.75% and 3.55% concentrations) after different time points show similar findings but with levels lower than 5 ng/mL at four hours (Huestis & Cone, 2004; Huestis et al., 1992). The average detection time for THC was 7.2 hours (for 1.75% THC) and 12.5 hours (for 3.55% THC) in the latter study. It should be noted that these aforementioned studies used blood serum rather than whole blood and that whole-blood THC concentrations would be lower.[1]

Several conclusions can be drawn from the above studies. First, the detection periods in blood are shortest for THC and longest for carboxy-THC (hydroxy-THC is in between the two). THC or carboxy-THC can be detected several days after abstinence, depending on the concentration cut-off chosen and a person's overall use. Very high readings of THC can be found in blood shortly after use (within the first half hour). Concentration then rapidly declines to a lower level that can be detected for several days after abstinence among daily users. Finally, heavy users will have longer detection periods than occasional users.

---

[1] Note: Positive drug tests can result from licit sources as well as illicit use. For example, a positive test for cannabis could be due to pharmaceutical THC (Sativex) (Cone & Huestis, 2007) or over-the -counter hemp oil products (U.S Department of Transportation, 2004).

Oral fluid—Oral fluid (less technically, called saliva) tests can be used to detect psychoactive components of drugs (i.e. typically THC for cannabis). A major issue with oral fluid tests for cannabis is contamination from residuals due to smoking. In other words, they can be positive for THC from traces left in the mouth that is not in their blood. In a study by Langel et al. (2013) with paired oral fluid and blood tests, numerous subjects had positive or high THC readings over 200 ng/mL in oral fluid but non-detectable amounts of THC in blood, possibly due to oral contamination. Presently, little is known about the length of time that surface cannabis residuals can remain in the mouth. Another drawback of oral fluid tests is that some individuals have a dry mouth and are unable to provide a sufficient saliva sample for analysis (Pil & Verstraete, 2008; Owusu-Bemphah, 2014).

In a study by Lee et al. (2012), 10 cannabis smokers each smoked a 6.8% cannabis cigarette weighing .79 g. Inclusion criteria included specific blood pressure and heart rate readings. Oral fluid THC levels were typically above 300 ng/mL in the first 15 minutes and over 100 ng/mL within the first hour after use, with a massive range in peak levels found between 68 and 10,284 ng/mL. All subjects had positive THC readings from oral fluid tests six hours after use, with a range of between 2.1 and 44.4 ng/mL. Clearly, this study found substantial variations in THC levels between subjects for the same time points after smoking. Similar results were found in other studies, as described below.

A nearly identical study by Milman et al. (2012) reported oral fluid THC concentration readings at different time points for 10 individuals administered a 6.8% THC cigarette. At .5 hours, four of 10 subjects were unable to provide a sufficient sample due to a dry mouth. At one hour, THC levels ranged between 35 and 1030 ng/mL—a massive degree of variability between individuals. At two hours, the median THC level was 31.7 ng/mL, with a large range between 11.6 and 1310 ng/mL. At four hours, when subjects would not be expected to be under the influence of cannabis, the median was 8.3 ng/mL with a range from 2.1 to 245 ng/mL. Clearly, the variability between subjects administered the same amount of cannabis is extremely high.

Toennes et al. (2010) administered doses of THC from 22 to 47 mg, intended to provide 500 ng THC per kg of body weight among 10 occasional and 10 chronic users. Chronic users had much higher THC peak concentration levels at both one and eight hours. Again, the variability between subjects in concentration levels was enormous at each time point. At one hour, THC levels ranged between 49 and 794 ng/mL in occasional users and between 24 and 1436 ng/mL in chronic users. The authors found that THC concentration averaged 8 ng/mL at eight hours post dose for occasional users and 16 ng/mL for chronic users. In addition, the authors reported a standard deviation of 7.4 for occasional users and 14.5 for chronic users at eight hours, which indicates very high variability among subjects.

This study found four of the 12 subjects had readings over 10 ng/mL at four hours. The authors also compared oral fluid and serum readings in a series of samples. They found that 24.1% of the negative serum samples tested positive when oral fluid samples were analysed. As found in other studies, the oral fluids to serum ratios were highly variable, with variations of ratios from .3 to 425 ng/mL. The authors concluded: "The large inter- and intra-individual variability observed precludes a reliable estimation of THC serum concentrations from oral fluid data…" (Toennes et al., 2010; p. 216).

Another study focused on detection times for THC in oral fluids (.3 ng/mL threshold) among frequent users (Andås et al., 2014). They found that negative samples could test positive at subsequent time points and that positive samples could be detected eight days after abstinence in one subject. Since a detection of THC was the main focus of this study, the findings are not useful for determining a possible cut-off for impairment.

These aforementioned studies were all based on confirmatory test results to detect THC. High average levels of THC were found in the first hour after smoking, followed by a gradual decline at subsequent time points. Extremely large variations between subjects in THC levels were observed in every study. Although confirmatory tests have been proposed by some as good biomarkers of impairment, persuasive research does not exist to show this is feasible with reasonable accuracy.

Correlations between blood and oral fluid THC—In addition to the controlled dosing studies above, studies have been conducted of paired oral fluid and blood samples from different subjects to assess the strength of correlations between the two. In the three cited studies below, the paired samples are independent of one another, meaning that no multiple observations of the same subjects were taken. There are other studies that report correlations from multiple paired measures of the same subjects at different time points, but this procedure is a violation of independent measures required by the statistical test. Three studies were found that met the inclusion criteria: Gjerde and Verstraete (2011), Langel et al. (2014) and Vindenes et al. (2012). These studies were conducted with sample sizes between 100 and 4,080 participants, and had very low coefficients of determination (r^2) of .12, .21 and .030, respectively. These findings demonstrate that oral fluid tests of THC cannot be considered a good substitute for THC blood tests.

Oral fluid tests are much more practical than blood tests to administer in a roadside situation and therefore they have received considerable interest in demonstrating their validity through comparisons to blood tests. In one paper, this desire to demonstrate validity has culminated in a violation of the purpose of statistics. Gjerde and Verstraete (2011), created of a statistical procedure, "prevalence regression," where percentiles between THC in oral fluids and blood were plotted and a new regression analysis was conducted with only nine data points. The r^2 value increased from .21 to .99, giving an illusion of a strong relationship. Essentially this method uses measures of central tendency (percentiles) that mask the reality of substantial variability between subjects in terms of oral fluid and blood ratios. In fact, if one uses this approach to create only two data points instead of nine, virtually any scatter plot of data, no matter how poor, can be turned into a perfect linear relationship (r^2 =1.0). No reference from a statistical expert was provided by the authors to support this approach. The authors did acknowledge large variations in oral fluid and blood ratios for THC and they should have just plotted the curve of best fit using basic regression methods. Their new statistical procedure of prevalence regression defeats the purpose of regression analysis as it essentially eliminates the variability between subjects, an important consideration for validity.

THC concentrations in oral fluid are much higher than in blood. The average conversion factor can be roughly estimated even with poor correlations. From the published studies, I estimate an average oral fluid/whole-blood ratio between 20 and 60, with higher ratios noted shortly after use (Hartman et al., 2015a; Langel et al., 2014; Vindenes et al., 2012).

Some authors have speculated that oral fluid samples can be analysed with gold standard confirmatory methods to draw conclusions of impairment. To date, use of oral fluids for this purpose is speculative and not supportable by existing science. An important finding of the controlled dosing studies cited is massive variations in THC concertation levels both between subjects and within subjects who were administered the same amounts of cannabis.

Point of collection for oral fluid tests—A major hurdle in testing is the development of a rapid, portable device that also can **accurately** measure the level of different drug compounds in the body. This task is not feasible with current confirmatory methods. A particular type of immunoassay screening test for drugs is the ***point-of-collection*** test, a simply administered drug test that produces findings of drug exposure within 10 minutes. There are about a dozen point-of-collection tests on the market for either urine or saliva samples, though there is considerable variability in the validity of these tests (Cone, 2001; Drummer, 2006; Blencowe et al., 2011). In a recent study, Blencowe et al. (2011) assessed the validity of the point-of-collection tests from oral fluids against confirmatory methods. The accuracy of these devices was particularly poor for cannabis at 73%, with average sensitivity of 38% and specificity of 95%.

The accuracy of point-of-collection tests has improved over the past few decades in relation to confirmatory tests. In a study on oral fluid point of collection for possible use in Canada, tests for cannabis (and other drugs) were validated against confirmatory tests of known drug users who acknowledged drug use in the previous three hours and known negative samples from police officers (Beirness and Smith, 2017). The accuracy of the point-of-collection tests was good at 92.3%, sensitivity (86.9%) and specificity (95.5%). An assessment of whether those with positive tests are impaired was outside the scope of this study.

Two oral fluid point-of-collection tests from a study by Beirness and Smith (2017) proposed by the Canadian government to screen drivers for cannabis were field tested with police officers (Keeping and Huggins, undated).[1] In this study, police described over 90% of tests as easy or very easy to administer, yet some aspects of the findings from this study are disturbing, suggesting use in the field may be fraught with limitations. Specifically, (1) 65% of test were administered outside the manufacturer's suggested operating temperatures, (2) three devices were found by police as defective, giving only positive results, and (3) officers had difficulty administering a swab for approximately 40% of the tests (Keeping and Huggins, undated). Clearly, there are potential pitfalls of using oral fluid tests in real-world conditions identified by this study and their proposed implementation by the Canadian government despite these limitations illustrates little concern for potential false positives.

When reviewing articles, it is important to keep in mind the purpose of the studies. If the purpose of the study is to detect prior drug use rather than impairment, some point-of-collection oral fluid tests have good accuracy. Most studies focused on comparison of oral fluid tests with confirmatory tests for the purpose of detecting use, and do not speak to the issue of performance deficits or impairment. Hence, we are still left with a poor understanding of the relationship between a higher cut-off level and cannabis impairment.

Urine—Urine tests, unlike blood and oral fluid tests, can be tampered with to give a false reading. Tampering is usually done with the intent to producing a false negative. Therefore, urine samples are typically first analysed to ensure the sample has not been adulterated. In terms of evading detection, there are a wide range of commercial products and adulterants that have been reported to produce false negatives (i.e. hide what would otherwise be a positive test). Jaffee et al. (2007) identified 12 commercial products that have been found to successfully mask marijuana, amphetamines, opioids, and cocaine or its metabolites. Simple household adulterants, such as bleach, vinegar, Drano, and detergents have been shown to be effective for producing false negatives. Most of the research has been conducted with initial immunoassay screening tests, since typically a

---

[1] The Drager DrugTest 5000 was selected as the official saliva test to be used by police.

confirmatory test is only requested for positive screening tests. Dilution by drinking large volumes of water or commercial products is somewhat effective; however, this can be detected with creatinine checks. Another method is urine substitution from clean donors. This can involve elaborate prosthetic devices and warming mechanisms to evade detection. To counter such evasion, direct observation of urination and testing samples for temperature, creatinine, specific gravity, pH, and oxidizing adulterants are recommended (Jaffee et al., 2007).

Urine tests are good specimens for identifying users of cannabis, as the detection periods for different compounds are longer than for blood. The window of detection for cannabis is highly variable and may be affected by numerous factors, including time and size of last dose, single dose versus multiple doses, volume of fluid intake prior to collection, kidney function (excretion of drug), liver function (metabolism of drug), as well as the kinetics of drug distribution, including volume of distribution and body fat. That being said, they are not feasible to assess impairment, because compounds are first detectable two to six hours after use (Swotinsky, 2015; p. 216) and therefore, those testing negative could be impaired if they used within a few hours before the test. Most cannabinoid compounds are detectable for very long periods after use. Carboxy-THC has been detected in urine for up to 60 days post-use (Verstraete, 2004). If the confirmation cut-off for carboxy-THC is 15 ng/mL, a threshold adopted by most workplace drug testing programs, it is unlikely an individual who used within a four-hour window would even test positive. In a study by Brenneisen et al. (2010), 12 male volunteers were administered a marijuana cigarette, then urine tests for different compounds were taken at various time intervals. Not a single subject had a reading above 15 ng/mL of carboxy-THC within a four-hour period of smoking. The authors note that urine has a wide detection window that "does not allow a conclusion to be drawn regarding the time of consumption or the impact on the physical performance" (Brenneisen et al., 2010; p. 2493).

In addition to carboxy-THC, both hydroxy-THC and THC can be found in urine samples, though these later two compounds have been infrequently studied in urine tests. In a study by Brenneisen et al. (2010), THC was detected in urine for all subjects at two hours, and one subject's at eight hours post-use. Urine samples to detect THC could have validity for use between two and eight hours, depending on the ng/mL cut-off. More research on detection of THC in urine, especially with daily and heavy users and mode of use is needed. Hydroxy-THC usually peaks between two and six hours after use. Overall, urine can be ruled out as a viable sample to assess performance deficits, regardless of the compound being detected, because it will produce false negatives within the two-hour period after use in occasional users.

Breath—Breath tests can be used to detect the presence of several drugs from exhaled air (Ossola, 2015). However, because the concentration levels of drugs in exhaled air are extremely low, the compounds can only be detected by mass spectrometry equipment. This equipment requires large amounts of space (similar to computers in the 1970s), which makes them highly unsuitable for field use.

Breath detection research is currently being conducted by companies such as Cannabix and Hound Labs to develop a breath test for THC compounds. Cannabix is working on a portable device using an ion spectrometry analysis. Cannabix reports that its device is aimed to target recent use within a two-hour window (Cannabix technologies). A portable test that could detect recent use within a two-hour window after last use could be useful for traffic safety; however, to date, no credible evidence exists of a breath test that can achieve this purpose.

CONCLUSION

Currently, the more accurate methods to assess cannabis impairment are with confirmatory analytic methods to detect THC in blood. Although some researchers have sought to detect multiple compounds or metabolites, none of this research has proven to be more accurate than THC detection alone to detect impairment. The important question is this: what cut-off should be used to best identify those who are impaired?

Comparisons can be drawn between the Breathalyzer for alcohol and blood tests for THC, which are summarized in **Table 6**. Blood is the gold standard for the purpose of assessing impairment from alcohol or cannabis. Breathalyzer tests for alcohol are an excellent substitute for blood tests and provide a close approximation of BACs. For THC levels, confirmatory tests between blood and all other types of biological are poorly correlated and cannot be substituted for blood tests. Confirmatory drug tests have excellent validity for detecting specific cannabinoid compounds, whereas screening tests are poorer in accuracy.

The Breathalyzer for alcohol has numerous advantages over blood tests for THC, as the Breathalyzer is less invasive, requires little training to administer and offers quicker results. Clearly, the Breathalyzer has many advantages over blood tests when considering practical issues of implementation in the field.

For cannabis, THC, the main psychoactive compound, has been the most thoroughly researched biomarker of impairment. Hydroxy-THC is a metabolite which produces stronger psychoactive effects, but has not been well studied. Carboxy-THC is readily detected in urine, but again, is not a viable biomarker for impairment due to its excessively long detection period. Generally, these compounds are detectable within different time periods after use, with the longest detection period for urine and the shortest detection period for blood tests. Future studies of THC in blood should report only whole-blood concentration levels because conversion levels between whole blood and serum/plasma are unreliable.

Table 6: *Comparison between the breathalyzer for alcohol and blood tests for THC*

	Breathalyzer (Alcohol)	Blood test (THC)
Accuracy compared with other biological samples	• The breathalyzer is an accurate substitute for a blood test to detect BAC. • Breathalyzers tests are excellent—$r^2 > .8$ for correlations of BAC by blood and breath (Zuba, 2008). • Oral fluid tests are also a good alternative to blood tests • Urine is not a suitable alternative to blood.	• Only blood tests are sufficiently accurate to detect THC concentration levels. • Oral fluids have a poor linear correlation with blood—r^2 values range from .03 to .12 of THC concentration in blood and oral fluids. • Urine is not a viable alternative to blood tests.
Practicality	• Non-invasive. • Requires little training. • Provides quick results.	• Invasive. • Requires training. • Requires laboratory methods with GC/MS for confirmatory tests.

Studies show that both confirmatory blood and oral fluid tests have similar patterns of concentration levels after smoking cannabis. THC levels spike within the first half hour after use, then quickly decline by about 90% at one hour, with a gradual reduction thereafter.

In this chapter, I have examined different detection approaches for alcohol and cannabis. In the next chapter, I examine the pharmacokinetics of alcohol and cannabis and in subsequent chapters, I examine the validity of the Breathalyzer for alcohol and blood tests for cannabis in terms of detecting impairment.

CHAPTER 4

Pharmacokinetics of alcohol and cannabis

PHARMACOKINETICS

Pharmacokinetics is a branch of pharmacology that focuses on the movement of drugs within the body and involves issues of absorption, distribution, metabolism and elimination of drugs in the human body. These three factors are related to the detection of a drug. This discipline is particularly useful for understanding the time course of drug concentrations in the blood after use, and for understanding the mechanisms for performance deficits. With this information, thresholds can be chosen to roughly map drug concentrations onto typical periods of being under the influence. These issues are best understood through controlled dosing studies. For a more technical and in-depth discussion of pharmacokinetics of cannabis, see Grotenhermen (2003).

Solubility

A major factor that influences absorption, distribution and elimination of substances and the choice of the best biological specimen is the properties of the substances themselves. Solubility refers to the ability of a substance to dissolve in a given solution. Solubility affects absorption and elimination rates in addition to determining which body compartments a substance will be distributed to. An important factor to consider is whether a substance is water-soluble or fat-soluble. Concentration levels in blood of water-soluble substances (i.e.

alcohol) correlate more closely with concentration levels found in the brain. On the other hand, fat-soluble substances such as THC have quite different concentration levels in the brain than in the blood. Collection of direct samples of brain tissue is not feasible in living persons, thus the best alternative of a biological specimen for chemical detection is blood serum.

Alcohol—Alcohol is water soluble, and as such, it is fairly evenly distributed in the blood, brain and other organs. This means that blood is a good indicator of the action of alcohol on the brain.

Cannabis—The main psychoactive component of cannabis is delta-9-tetrahydrocannabinol (THC). THC is a fat-soluble molecule that preferentially distributes to fatty tissues and fatty organs, such as the brain and liver. High-perfusion tissues, including the liver, will absorb THC at a much quicker rate (Ashton, 2001). THC is then released back into the blood at a lower and largely unknown rate. THC will also be slowly distributed to and redistributed from low-perfusion tissues like skeletal muscle. Only one study has compared THC levels in the brain and blood; however, this was performed in 10 deceased individuals (Mura, 2005). Re-analysis of the coefficient of determination showed this relationship between blood THC and brain THC was poor (r^2=.19). This finding suggests that blood samples of THC may not be an accurate measure of impairment.

Absorption and distribution

Absorption refers to how quickly a substance is absorbed into the body and distribution refers to how substances migrate to different body parts and into cells or across tissues and organs, and in particular the brain. Biological tests for alcohol and cannabis are indirect measures of the influence they have on the brain, as direct tests of the brain are not feasible. These indirect measures are associated with error due to various factors. One influential factor is how drugs are consumed: whether they are taken orally, intravenously or inhaled.

Alcohol—Alcohol, which is only ingested orally, is quickly absorbed from the stomach into the blood. It reaches ***equilibrium*** between the blood and brain compartments in the first half hour after drinking and remains fairly constant thereafter.

Smoking cannabis—After smoking cannabis, THC is rapidly distributed from the lungs to blood (see **Figure 1**). It then travels to the brain, fat and perfusion tissues, as well as the liver. The liver produces two main metabolites (by-products of the metabolism): first is hydroxy-THC, a psychoactive substance, and second is carboxy-THC, a non-psychoactive substance. The majority of these metabolites are excreted in urine along with small amounts of THC (Brenneisen, 2010). What remains is redistributed in the bloodstream. THC levels rise very rapidly in blood after smoking, demonstrating rapid absorption, peaking within three to 10 minutes (Hunault et al., 2014; Grotenhermen, 2003). In one study, the range in the peak amount of THC in the high dose condition was between 76 – 262 ng/mL in plasma (Huestis et al., 1992). When cannabis is smoked, hydroxy-THC peaks later in the blood at concentration levels of about 3–15% of THC levels. The degree to which hydroxy-THC or the interaction of THC with hydroxy-THC creates additional performance deficits is unknown. There is a major difference in THC concentrations after smoking cannabis compared to eating it.

Figure 1: *THC distribution in humans from smoking cannabis*

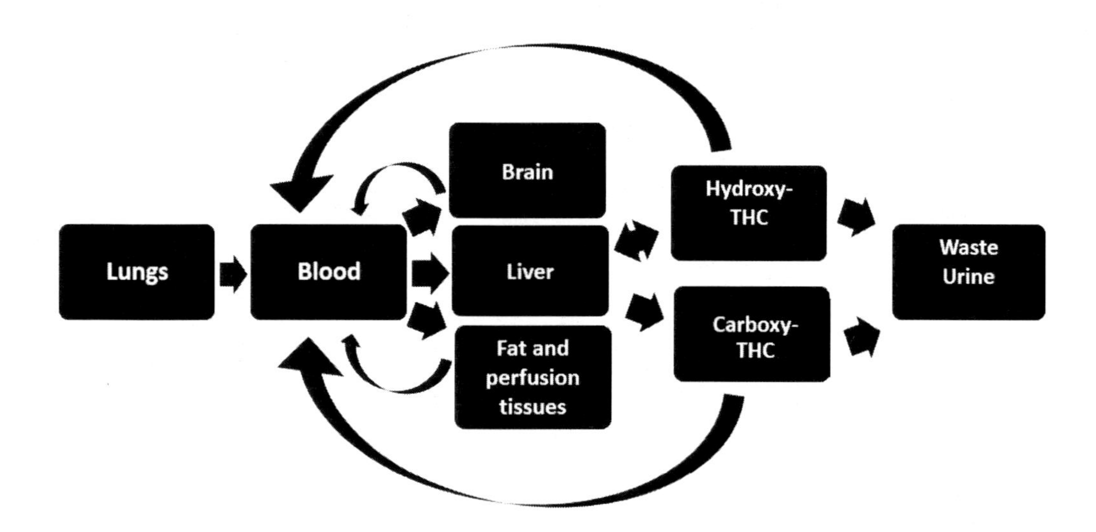

Ingesting cannabis—When eaten, absorption of cannabis is much slower than after smoking (Kapur, 2009; Sharma et al., 2012), and distribution is much different, as cannabis is first absorbed by the stomach rather than the lungs (see **Figure 2**). After it enters the stomach and blood, most of the drug then passes through the liver where it is metabolized and some THC passes into the blood, where it is distributed to other body parts including the brain. THC rapidly penetrates fat tissues where it is slowly redistributed back into the blood. Some of the THC is not absorbed and much of it is metabolized first into hydroxy-THC and second into carboxy-THC. Maximum THC levels are typically reached between the first and second hour, and it is possible that multiple peaks can be reached for some people (Sharma et al., 2012). These compounds then are distributed to the blood and brain. Peak

THC concentration levels are typically low compared to smoking—about 5 to 10% of the peak level when smoked. However, levels of hydroxy-THC are much higher when eaten than smoked, often exceeding THC concentration levels (Ménétrey et al., 2005). The ratio of THC to hydroxy-THC is much greater when cannabis is ingested versus smoked. We do know that hydroxy-THC is psychoactive. More research is needed to understand the pharmacokinetics of ingested cannabis.

Figure 2: *THC distribution in humans from ingesting cannabis*

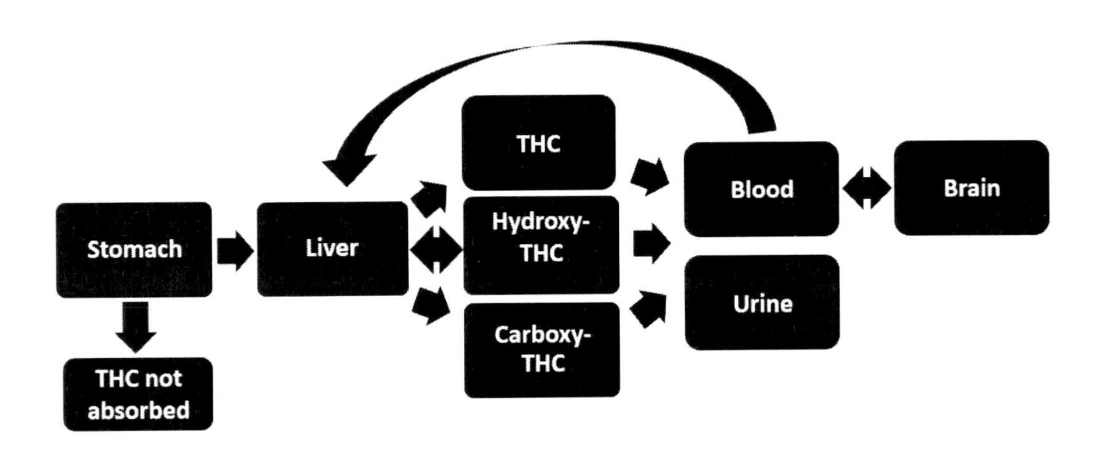

Elimination

Elimination is any process by which a drug is excreted from the body. This process is related to metabolism, mainly in the liver, and detection of drug concentrations. Individual differences in physiological characteristics, diet and food intake affect both absorption and elimination. As previously noted, fat-solubility strongly affects elimination; fat-soluble drugs can remain in the body for days, whereas water-soluble drugs are eliminated quickly and at a more consistent rate.

Alcohol—Several variables can affect the rate at which alcohol is eliminated from the body. Erik Widmark, a Swedish pioneer in the pharmacokinetics of ethanol (absolute alcohol) in the 1920s, first developed a method for measuring alcohol in the blood (Lerner, 2012). Although any given person metabolizes alcohol at a fairly constant rate, the elimination rate between individuals varies based on individual characteristics. Men generally metabolize alcohol faster than women and younger people sooner than older people. Elimination rates are higher among heavier drinkers and lower among lighter drinkers (Kapur, 2009). Wigmore (2014) and Zakhari (2006) reported that alcohol elimination rates vary approximately 3-fold among individuals. Widmark also noted that eating food when drinking considerably lowered BAC level. Although alcohol elimination does vary among individuals, the more important issue is the degree of variation in performance between subjects at the same BAC. Early studies have found that performance at the same BAC level is worse when BACs are trending upwards rather than downwards (Jones and Vega, 1972). Extensive research shows that ***back-extrapolation*** of current BAC and a few key variables produces good estimates drawn on the amount of alcohol in a person at an earlier point in time.

Considerable research has been conducted to assess the potential degree of error for calculating the amount of alcohol in a person based on BAC and other measurements. Searle (2015) identified six potential sources of error and noted that the amount of error in estimates should be calculated on a case-by-case basis. ***Coefficients of variation (CV),*** a formula for expressing variation in samples that is useful for comparing measures with different units of analysis, have been estimated by several authors. CV values below 15% for estimations of drink volume based on BAC and other individual factors suggest a fairly high degree of accuracy. Breathalyzers in North America are calibrated to produce lower BACs in order to compensate for these errors.

Numerous BAC calculators available on the internet provide estimates of BAC. One produced by the Alcohol Help Center includes five variables: sex, country (definitions of standard drinks vary among countries), weight, hours since having the first drink and number of standard drinks. I used this program to estimate the rate of absorption and elimination of alcohol for a 170-pound Canadian male consuming five drinks within one hour after the first drink. The approximate BAC level at different time points is plotted in **Figure 3** (below), along with similar data for THC concentrations in blood. This figure shows that a person drinking this amount of alcohol will have a BAC of about .04% alcohol after four hours. The rate of elimination is constant and it would take a full eight hours to obtain a BAC level of zero.

Cannabis—Little research has been conducted regarding estimations of the amount of cannabis used based on blood concentration levels. In one study, subjects who received larger doses of smoked cannabis had marginally higher levels of THC in their blood than those who received low doses (Hunault et al., 2014). A high degree of variability exists between people for THC blood concentration levels at different time points, given the same amounts used. There are findings of CV values for THC blood concentrations at different time points after subjects have been administered the same doses of THC. In a study by

Hartman et al. (2015a), CV values were presented at four time points (up to 2.3 hours after smoking) for 19 subjects. CV values ranged from 46% to 82%, which indicates above average dispersion. This degree of variability is too large for back-extrapolation (Hartman et al., 2016). A peak level can vary substantially, often between 50 to 100 ng/mL in whole blood, depending on the person and dose. Peak levels will be even higher in samples of plasma or serum.

Formulas for the relationship between smoked cannabis and blood THC have been proposed but not independently validated. For example, Sticht and Kaferstein (1998) (reported in Grotenhermen et al., 2007) suggest that a 70 kg male who consumes 19 mg THC will have a blood serum concentration of 4.9 ng/mL (CI 3.1–7.7) after three hours. Extrapolation of this finding to whole blood, to females, other body weights, and other time points is problematic. As pointed out in this book, substantial variation exists between individuals for blood THC who are administered the same amount of cannabis. This formula must therefore be treated as speculative, as research validating this work, or that generalizes this formula to those with other conditions (i.e. sex, weight, metabolism etc.), has not been adequately established.

Figure 3: *Absorption and elimination of THC and alcohol*

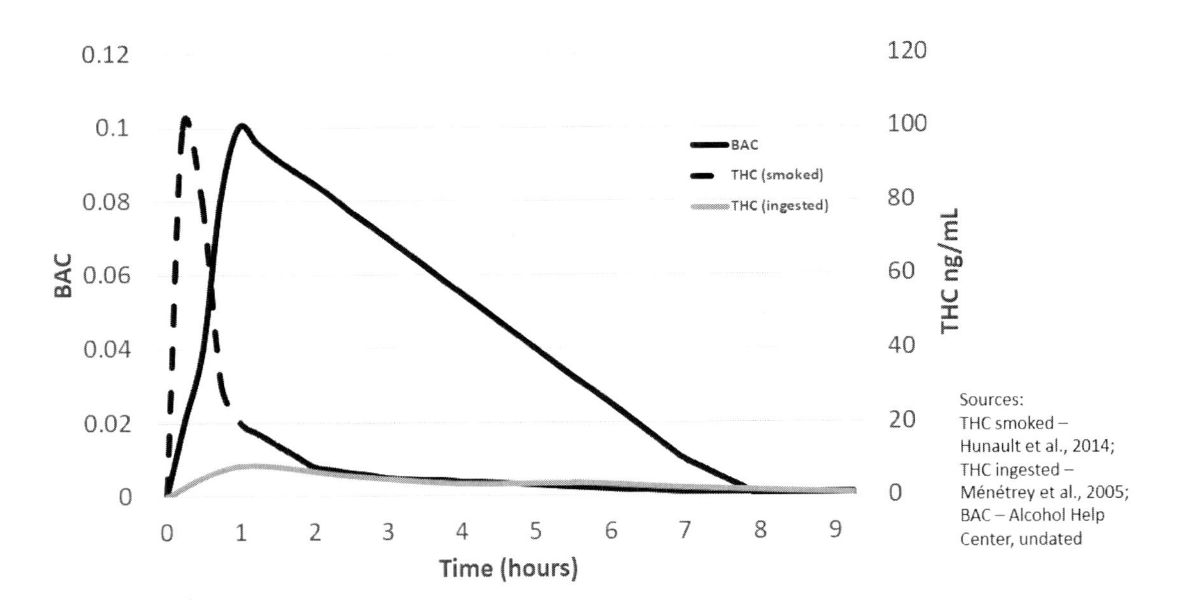

Figure 3 (above) compares the absorption and elimination of smoked and ingested THC with alcohol. This figure is adapted from serum to whole blood (using a very

approximate conversion factor of 2/1) presented by Hunault et al. (2014). After smoking 50 mg THC of cannabis, on average THC levels rise very rapidly in blood and peak within three to 10 minutes. Around an hour after smoking, THC levels typically decline by about 90%, and then very gradually decrease. In conclusion, the elimination rate of THC from the blood is **not** consistent over time. Elimination is rapid shortly after use, followed by a slow decrease as THC is re-released from fatty cells and metabolized.

The numbers for cannabis ingestion are based on a study by Ménétrey et al. (2005) for subjects who ingested at 45.7 mg dose of THC. In this situation, the average THC levels were very low in the first hour after eating (the highest mean concentration was 8.4 ng/mL in whole blood) compared to smoking, and subjects on average experienced gradual reductions to 3.5 ng/mL after 5.5 hours. By contrast for alcohol, BACs rise rapidly and then decline at a constant rate.

Most of cannabis, regardless of how it is taken (smoked or ingested), is eliminated from the body within five days in the form of two metabolites—hydroxy-THC and carboxy-THC (Sharma et al., 2012). Hydroxy-THC is eliminated more in feces, whereas carboxy-THC is eliminated more in urine (Sharma et al., 2012).

CONCLUSION

A chief pharmacokinetic difference between alcohol and cannabis is that alcohol is water-soluble, whereas THC is fat-soluble. The main implication of this difference is that alcohol in the blood has a strong relationship to alcohol in the brain, whereas for cannabis, the relationship between blood THC and THC in the brain is only strong shortly after use. Another finding is that individuals eliminate alcohol from the blood at a constant rate while THC is eliminated extremely rapidly from the blood shortly after smoking, followed by a period of slower metabolism. Variations between individuals in the elimination of THC are much greater for cannabis than alcohol.

We have a good evidence-based understanding of the major variables that affect BAC after ingesting alcohol. For cannabis on the other hand, our knowledge is still poor. We know that alcohol is completely eliminated from the body within a much shorter time frame than blood THC. The difference is especially great for daily users of alcohol versus regular cannabis users. Daily users of alcohol eliminate alcohol at a quicker rate than occasional users; however, daily users of cannabis have THC in the blood for much longer periods of time than occasional users. A critical question of importance is the strength of the relationships between these biological measures and performance deficits.

CHAPTER 5

Substance use and performance

In this chapter, alcohol and cannabis are examined in terms of their acute (i.e. immediate) performance deficits. The studies reviewed in this chapter include paired measures of performance and biological tests (either the Breathalyzer for alcohol or blood tests for cannabis) to assess the nature and strengths of their relationship. In order to understand the relationship between THC blood tests and performance deficits, direct measures of both performance deficits and THC blood concentrations are needed. One aim will be to identify the period of time after use when an individual is most likely to have performance deficits equivalent to being impaired by alcohol.

How do we assess the accuracy of drug tests for impairment? The best studies first determine a critical and meaningful criterion for performance. Second, paired measures of drug concentration and performance level are taken.

I define impairment broadly as meaningful decrements in psychomotor or other perceptual or cognitive functioning. Impairments are shown by research to be related to increased risks of accidents or injuries, such as those that could occur from driving a motor vehicle. The most common form of impairment and the reason that people use drugs is based on the immediate or *acute* effects. These immediate effects are the focus of impaired driving laws.

MEASURING THE AMOUNT OF ALCOHOL AND CANNABIS CONSUMED

Despite the fact that international units of analysis for alcohol are currently defined as grams of ethanol or litres of absolute alcohol, conveying the concept of a standard drink has been an uphill battle for both researchers and legislators. In Canada, a standard drink is a 355 mL (12 ounce) beer at 5% alcohol, a 146 mL (5 ounce) wine at 12% alcohol, or 44 mL

(1.5 ounce) of spirits at 40% alcohol (Canadian Public Health Association, 2006). However, research shows that people often do not consume alcohol in standard drinks. Wine and spirits are often free poured and beers come in a variety of sizes and alcohol contents that do not conform to standard drinks. Also, the definition of a standard drink varies greatly among countries, as do amounts specified by recommended safe drinking guidelines (International Alliance for Responsible Drinking, 2017).

The challenges faced in terms of providing international standards are exacerbated with cannabis and other illegal drugs. Cannabis comes in a variety of different concentration levels, estimated to average about 8.5% THC in cigarettes analyzed in 2008 (McLaren et al., 2008). Potency levels have been trending upward over time. In one experimental study, subjects were administered THC cigarettes of 23% potency (Hunault et al., 2014). Cannabis also comes in the form of hash, oil and synthetic products (Sativex). It can be smoked in joints of various sizes, through a pipe or vapourizer, or eaten. ***Titration*** of inhaled products in ***experiments*** has foiled some researchers (Hartman et al., 2015a), making it challenging to carry out a rigidly controlled study. Researchers have attempted to address some of these variations for determining dose amounts (Zeisser et al., 2008); however, much more investigation needs to be done.

The best way to define cannabis is in milligrams of THC, because it combines both the percentage of THC and weight of the product into one measure. A standard single dose of cannabis commonly adopted for medical marijuana is 10 mg THC, although there are no uniform dosing schedules (Government of Canada, 2016). An average joint weighs about .5 gm or 500 mg. A 500 mg joint of 12% THC would have approximately 60 mg of THC, which would be considered a high dose if smoked wholly by one person, and a very high dose if eaten. A moderate amount smoked is about 40 mg THC, and a small amount is about 20 mg THC. I support adopting an initial standard for a dose of cannabis of 10 mg THC as it will be simple for the public to understand.

Laboratory studies

Strengths and limitations

Laboratory studies are conducted in controlled environments and are most useful for identifying the precise nature of performance deficits produced by specific drugs. The best research design typically consists of a ***double-blind*** experimental approach where subjects are ***randomly assigned*** to a group administered a drug or to a placebo (i.e. no pharmaceutical effect) group. Ideally, two identical groups of people living in a single environment are given two clearly different treatments, and then monitored for the possible differences of effects. This type of study has an experimental and control group that are assessed in terms of performance differences. This type of study has strong ***internal validity***—that is, the study design allows for strong inferences of cause and effect because the groups are equivalent in all other respects, except for the intervention. This type of validity should not be confused with validity of a drug test that was described in Chapter 2.

The main limitation of laboratory and simulator experiments relate to ***external validity*** issues because drug use and the driving environment in experimental designs may not reflect actual patterns of drug use and driving in the real world (Del Rio & Alverez, 1995a; van Laar et al., 1993; Volkerts et al., 1993). Performance deficits found in laboratory conditions may not translate into safety concerns in the real world for several reasons. For one, those who use cannabis may be unlikely to drive a car, possibly because the drug has a risk-aversive influence. Alternatively, the effects may not be substantial enough to meaningfully increase safety risk, or because individuals compensate for their condition. Poor generalizability of laboratory research to real-world conditions can occur for other reasons, such as the use of healthy volunteers (naïve or occasional users versus daily drug users), or duration and dosages of drugs that are different from how they are normally consumed (de Gier & Laurell, 1992; Del Rio & Alverez, 1995a). In fact, there is some evidence for both alcohol and cannabis that findings in laboratory settings may not translate into real-world conditions.

Determination of performance thresholds

Given these limitations, laboratory approaches have been proposed to assess how different types of performance deficits might translate into elevated crash risk in real-life situations that could be classified as impairment. Five key tasks were identified by a panel of experts in the field (Kay & Logan, 2011):

(1) alertness/arousal
(2) attention and processing speed
(3) reaction time/psychomotor functions
(4) sensory-perceptual functions
(5) executive functions

The panel indicated that a psychoactive substance that "impairs performance in any of these domains at a magnitude known to be associated with increased crash risk is presumed to have a negative impact on driving safety" (Kay & Logan, 2011; p. 6). However, the report did not go further and provide an ***operational*** definition of these thresholds, meaning there was no clear explanation of how to best measure these deficits. The identification of these domains provides a useful guide on the types of deficits that should be included in laboratory studies of alcohol and drug effects. However, the extent to which each of these domains relates to crash risk is largely unaddressed. Some domains will likely have a more detrimental impact on traffic safety than other domains. For example, poor psychomotor functions that cause drivers to weave outside their driving lanes is more detrimental to driving safety than slow reaction times that cause drivers to drive twice as far as normally necessary behind another car. Unfortunately, there has yet to be a proper analysis of the extent to which each of these domains is related to crashes, and performance cut-offs have not been proposed. One issue is the challenge of deciding a threshold when these constructs are typically measured as continuous variables. The value of validity studies is that optimal thresholds with a balance of sensitivity and specificity can be determined.

Another approach used to assess the degree of deficits from cannabis has been to draw direct comparisons of deficits from alcohol to those from cannabis. This is accomplished by having subjects complete performance tests under both an alcohol and cannabis condition. The amount of cannabis required for deficits equivalent to those for alcohol at a particular BAC (often .05% or .08%) can be determined. One study, described below, used this approach (Hartman et al., 2015b).

Acute symptoms and duration of impairment

Alcohol

The acute symptoms of alcohol are dose-dependent with symptoms such as increased talkativeness and sociability at low consumption levels (typically one or two drinks, <.05% BAC, depending on a person's weight and other characteristics), and symptoms of incoherent speech, staggering and memory loss at high consumption levels (typically five or more drinks, >.08% BAC). Alcohol at moderate levels (.05 BAC) can impair both psychomotor skills and divide attention, which increases the likelihood of accidents when operating complex machinery or driving a car. Tracking performance, which requires hand and eye coordination associated with safe driving is sensitive to the effects of alcohol (Martin, 1998). There are individual differences in deficits caused by alcohol, related to tolerance and the ascending or descending BAC (Martin, 1998).

The duration of performance deficits depends on the amount of alcohol consumed or how long a person's BAC is above a particular threshold. The vast majority of laboratory-style studies on performance deficits from alcohol have paired BAC levels with performance indicators. An important research question is to establish a threshold of performance deficits whereby most individuals could be classified as impaired. Based on recommendations by Fell and Voas (2014), .05% BAC is a reasonable cut-off.

Cannabis

Most of our knowledge of the acute effects of cannabis is based on self-reports by users and laboratory studies. Main symptoms of intoxication include increased drowsiness, impaired short-term memory and decrease in attention. Euphoria, feelings of heightened perception, increased appetite and distortion of time are also common (Hall and Solowij, 1998). Numerous laboratory-style studies and reviews have been conducted on the acute effects of cannabis on performance. A review article published over three decades ago provided an assessment of the acute effects of cannabis on driving, which have not been refuted by subsequent research (Moskowitz, 1985). Coordination, tracking, perception and vigilance were all negatively affected by cannabis. Limitations noted by this author are that a clear determination of the magnitude and nature of psychomotor performance is unknown, and studies of heavy chronic users are lacking. Unfortunately, these same limitations could be said to be present in existing research today. Several dimensions of performance were examined in the 1985 review: coordination, reaction time, tracking, as well as sensory and perceptual functions. A recent review article concluded that cognitive deficits are associated with acute effects, but fewer studies show motor deficits similar to those found for alcohol (Prashad & Filbey, 2017). However, a review paper by Bondallaz et al. (2016) found evidence of increased lane weaving, as well as longer distances behind vehicles, under cannabis conditions. Another systematic review (Broyd et al., 2016) reports that the wide ***heterogeneity*** of studies (in terms of cannabis exposure, samples and measures) makes it challenging to draw meaningful conclusions, as the studies are very different from one another. These authors evaluated several cognitive domains and found strong scientific evidence that the acute effects of cannabis lead to deficits in verbal learning and memory, attention and psychomotor function. The overall research shows that cannabis causes performance deficits; however, the establishment of performance thresholds that could constitute impairment similar to the effects of a BAC of .05% has not been properly addressed.

Generally, based on subjective effects, a peak psychoactive high is experienced after smoking cannabis within 30 minutes, which lasts approximately two hours with a return to normal within four hours (Grotenhermen, 2003; Raemakers et al., 2004; Kelly et al., 2004). Research shows that that a high for smoked cannabis is nearly certain within a one-hour period if smoked, and the majority of people will not be experiencing any major deficits after four hours. A shorter average duration of 45 minutes for being high was noted in one study for a 9 mg THC cannabis cigarette (Harder and Rietbrock, 1997). After eating the drug, a peak high is typically noted between one to three hours, which declines to low levels within six hours (Grotenhermen, 2003).

In 2005, a panel of experts on driving under the influence of cannabis made recommendations for objective criteria to determine impairment by cannabis (Grotenhermen et al., 2005). Their analysis found maximum performance deficits in 64% of test results at 20 to 40 minutes after smoking, which compares to 35 to 40% of test results showing any performance deficits between one and 2.5 hours after smoking. Clearly, dosage and time period after use are very important variables in terms of likely deficits, but variability between individuals is so great, making it difficult to draw reasonably precise conclusions on how these variables might be validly related to impairment. A bold finding of the subsequent published paper was, "A comparison of meta-analyses of experimental studies on the impairment of driving-relevant skills by alcohol or cannabis suggests that a THC concentration in the serum of 7–10 ng/mL is correlated with an impairment comparable to that caused by a blood alcohol concentration (BAC) of 0.05%" (Grotenhermen et al., 2007, p. 1910). They suggest this concentration level is equivalent to 3.5 to 5 ng/mL in whole blood. The main conclusion from this study is not justified by the scientific methods. Specifically, the conclusion is based primarily on laboratory studies where no blood tests were conducted, as acknowledged by the authors. Conclusion about an association of blood tests for THC and impairment cannot be justified unless direct comparisons are conducted between the two—and no studies are cited where these comparisons were done. A common principle of research is that one should not draw inferences about findings (i.e. generalizations) to a different population than from the studies which the participants of the study were selected.

The above studies have several notable limitations. We do not have a clear idea of the amount of cannabis needed to produce impairment, how long impairments may last and how to measure impairment. Although we can provide some general conclusions based on averages, variations among individuals is large. For the laboratory studies, performance tests were usually performed within the first hour after cannabis administration (Grotenhermen et al., 2005). It appears that performance deficits can meet a threshold of impairment at some dosage and time point after use, but we do not have a good idea **when** this occurs after use for a given individual. One difficulty is related to the failure of studies to make any distinction between performance deficits and impairment. There are some studies that compare biological test measures with performance that are reviewed in greater depth later in this chapter.

STUDIES ON THE RELATIONSHIP BETWEEN PERFORMANCE DEFICITS AND BIOLOGICAL TESTS

In "The purpose of biological tests" section of Chapter 2, I described the epidemiological process of assessing the validity of diagnostic tests to assess impairment based on laboratory findings. Studies on how researchers applied this method for alcohol and cannabis impairment were examined. This ideal process involves the following steps:

- Create a gold standard to define cannabis impairment in terms of performance deficits is dangerous for nearly all people.
- Conduct paired biological drug tests with the gold standard of impairment to compare the average drug concentration levels (THC) for those impaired and non-impaired.
- Determine the best cut-off that reflects an acceptable trade-off between sensitivity and specificity for impairment.
- Calculate Kappa, which ideally should be over .75 for excellent agreement.

Unfortunately, studies that used these ideal methods were not found. Rather, laboratory studies typically assess the relationship between BAC levels and performance deficits as continuous variables. Although such studies provide a good foundation on the nature and magnitude of the relationship between performance deficits and substance concentrations, they are less helpful for arriving at an evidence-based per se limit for impairment.

Performance deficits at low levels

Alcohol

Performance deficits have been detected at low levels (< .02% alcohol) (Ogden & Moskowitz, 2004), but how driving is affected at these levels in the real world is largely unknown. One important issue is understanding the meaning of these deficits at these levels in terms of crash risk and how these conclusions were derived. In an earlier review paper on the same topic, Moskowitz and Florentino (2000, abstract) concluded that "alcohol impairs some driving skills beginning at any significant departure from zero." However, when examining a summary of behavioural test results from studies with BACs between .02 and .029%, about five times as many studies reported no impairment compared with impairment at this range. As well, no information is provided on whether there were performance improvements at this range. Often authors don't even test for possible improvements at this range, a procedure known as ***one-tailed statistical test***, common in some disciplines, such as psychology (the discipline of Herbert Moskowitz), but rarely used in other disciplines, including epidemiology. Epidemiologists argue that one-tailed statistical tests can obscure important findings. In any case, conclusions from such studies do not reflect the preponderance of the research and are likely not relevant to the real world.

The majority of research has found a strong negative linear relationship between BAC level and performance, meaning that as BAC levels increase, performance decreases and that the relationship is linear. For example, Dawson and Reid (1997) found a strong linear relationship between the two ($p<.05$, $r^2=.69$). Although the vast majority of studies have demonstrated a strong relationship between BAC levels and performance, none were found that examined the validity of a particular cut-off of .02% for impairment. Data from existing research can be misleading. For example, in the study by Dawson and Reid (1997), drivers at less than .02% alcohol performed better than they did at baseline. Drivers at .05% alcohol performed at about 96% of baseline and those at .08% alcohol performed at about 90% of baseline. Such data does not support deficits at .02% alcohol as no performance deficits were found at this level. Studies such as these do not support a threshold because data is presented as a continuous measure and performance deficits that constitute impairment are not operationally defined. Also, this study and many others on BAC level and impairment did not classify subjects in terms of sensitivity and specificity by choosing an optimal cut-off to assess validity. In another study, performance improvement for throwing darts was found at .02% BAC but deficits occurred at .05% (Reilly & Scott, 1993). Such studies illustrate the limitations of describing performance at these levels as impairment. Studies on low BAC cut-offs and impairment did not classify subjects in terms of sensitivity and specificity as I recommend. The vast majority of studies have demonstrated a strong relationship between BAC levels and performance but have not shown performance deficits are meaningful at low levels. BACs of .08% do represent substantial and meaningful declines in performance.

Cannabis

Like alcohol, cannabis produces cognitive and behavioural deficits in a dose-related manner (Hall and Solowij, 1998). However, research is too limited to draw any conclusion on how low doses of cannabis affect performance. At low doses of cannabis (defined as nine to 18 mg THC) a review study did not find performance deficits between 2.5 and 12 hours after use (Grotenhermen et al., 2005). Given the difficulty of researchers to provide a meaningful operational definition of cannabis impairment, mixed findings of many studies on the acute effects of cannabis in terms of performance deficits, and the weak validity of DRE assessments for recent use, low doses do not appear to produce meaningful performance deficits.

A question of importance is: what is a low dose and what THC levels in blood can be found at low doses? In fact, there is no accepted definition of a low dose, and the effect of cannabis on one individual may be quite different from that of another. Microdosing cannabis (taking the lowest dose possible to achieve a particular effect) has increased in popularity, especially for medical issues, such as anxiety. The extent of microdosing in the general population is largely unknown. In terms of THC levels in blood, Hunault et al. (2014) described a low dose at 29 mg THC, and found blood p levels peaked on average at about 150 ng/mL THC (very roughly equivalent to about 75 ng/mL in whole blood) and after 2 hours, THC levels appeared only slightly higher in high-dose (69 mg THC) conditions. Overall, the research is sparse in terms of how low doses affect performance but THC levels do appear to get fairly high (about 75 ng/mL) at low doses.

THC in blood

In a review of the literature, I found two review papers that made recommendations of THC blood level ranges associated with impairment that deserve special attention (Grotenhemen et al., 2007; Hartman & Huestis, 2013). Below I focus on the assessments of laboratory studies of THC doses, performance and THC blood tests.

Grotenhemen et al. (2007)—Grotenhemen et al. (2007) proposed a possible per se limit of 7–10 ng/mL THC in serum, which they indicate is equivalent to 3.5–5 ng/mL THC in whole blood. This expert panel reviewed both laboratory studies on cannabis and performance and epidemiological studies of cannabis and accident risk and weighed the laboratory studies heavily in arriving at their conclusion above. In their review, they found 87 studies of cannabis administration and a variety of performance tests. These studies provide tangible evidence of performance deficits from the acute effects of cannabis, summarized earlier in this chapter. Unfortunately, few of these studies included blood tests for THC so that direct measures of blood THC with performance could be assessed. Instead, the authors estimated blood THC concentrations of subjects in these studies based on the dose, and other factors (weight and sex were noted). The authors did acknowledge substantial within- and between-subject variability in THC blood levels over time and that this was a limitation of their approach. In my opinion, this approach is scientifically unsound and the conclusions above cannot be justified by these methods.

Hartman and Huestis (2013)—These authors conclude that recent smoking of cannabis and 2–5 ng/mL THC in whole blood are associated with substantial driving impairment based on a review of both laboratory and epidemiological observational studies. This paper did not attempt to determine THC blood levels from the laboratory studies reviewed. No findings from any laboratory study with paired blood THC and performance measures was reported in this review. Therefore, it must be assumed that the laboratory studies refer to the acute effects of recent smoking and not impairment related to THC from blood tests. Some empirical studies that examined the relationship between paired measures of THC in blood and performance are reviewed below.

Figure 4: *The relationship between THC concentrations in whole blood and subjective feelings of being high for* ***smoked cannabis*** *of 69 mg THC*

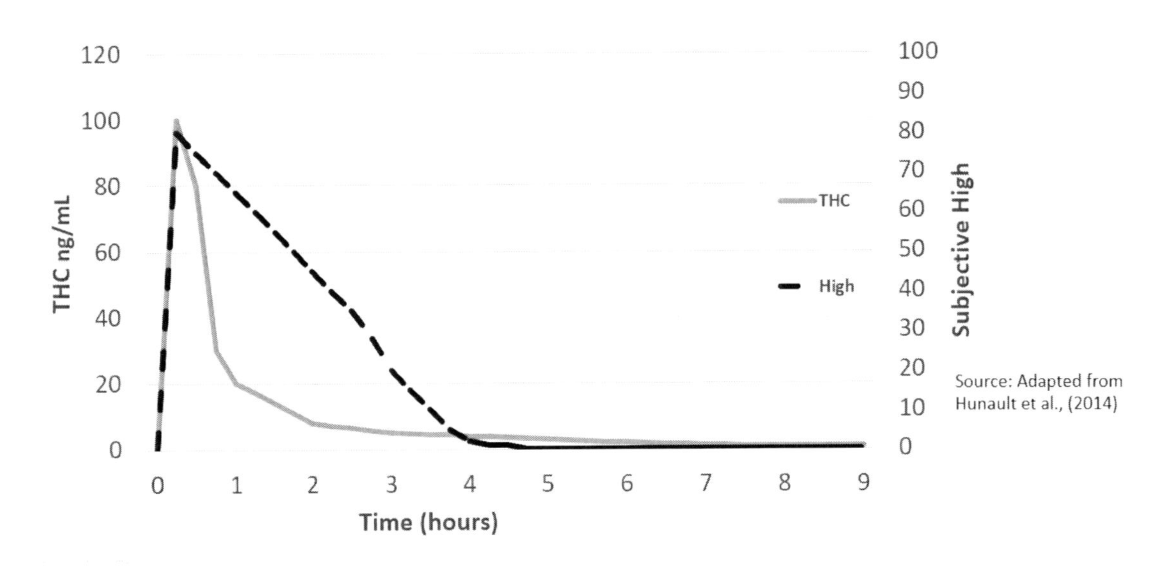

Hunault et al. (2014) – In this study, the relationship between three different THC potency cigarettes (from 29 mg to 69 mg THC) and subjective reports at various time points were examined among 24 recreational users. **Figure 4** shows a comparison of the THC[1] levels and levels and subjective levels of intoxication with a 69 mg cannabis cigarette. The peak high and peak THC concentration levels occurred at a similar time at under 30 minutes

[1] Hunault et al. (2014) assessed THC with serum. Figure 4 has been adapted to whole blood levels from serum using a conversion factor of .5.

after smoking. However, THC level declines much faster than the decline in feeling high, an observation noted by Moskowitz over 30 years ago (Moskowitz, 1985). It is clear that a relationship exists between THC levels and subjective high; however, the relationship is strong within the first hour after smoking, and becomes much weaker in the next hour and very weak afterwards. The steep decline in THC concentration levels level off at about 10–20 ng/mL, followed by a gradual decline in THC concentrations. For subjective high, the peak high is achieved shortly after smoking and a linear decline occurs over four hours. Selection of a THC cut-off within the first hour after use of over 10 ng/mL during the steep rise and decline will have better overall validity for impairment than a lower cut-off within an area of a more gradual decline. Overall, higher thresholds will yield better specificity and lower sensitivity. The best overall accuracy of THC in blood for impairment, based on this study may be in the range of 10–20 ng/mL, and can be established through analysis of the raw data with an approach known as ***Receiver Operating Characteristics (ROC).*** This approach is best when false positives and false negatives have equal importance and with a gold standard for impairment.

Although this study has several merits, summary statistics are used to present findings. While summary statistics are useful for showing average conditions, measures of central tendency tend to obscure variations between subjects, known to be extreme (Moskowitz, 1985). Individual data is needed to properly assess the validity of a particular THC cut-off in terms of impairment. Moreover, from this data, we do not know what level of subjective high could constitute impairment. Knowledge of the variability between individuals and a good benchmark for performance deficits that constitute impairment are needed in order to properly assess the validity of a given cut-off for impairment, which was not the purpose of the paper by Hunault et al. (2014). This purpose is rarely found in the literature, which means that cut-offs for cannabis impairment are based on a weak scientific foundation. In order to achieve this goal, individual level data is required, not descriptive summary statistics. The approach used by Hunault et al. (2014) is not a substitute for the methods to assess validity described Chapter 2, but provides some empirical evidence of where to choose a valid cut-off.

Hartman et al. (2015)—This study has a randomized, double-blind, and placebo-controlled research design—a classic experimental approach. The outcome measures were lateral control, such as weaving, in a very realistic driving simulator. They were also able to calibrate driving performance with blood THC levels in cannabis through comparison to alcohol-impaired driving. They noted that 52.6% of their subjects showed evidence of titration (or a subject inhaling more or less cannabis to obtain the optimal dose), which may help explain the substantial variations between subjects who were administered cannabis in other studies. The authors clearly found deficits related to THC blood levels. They concluded that blood THC concentrations of 8.2 ng/mL and 13.1 ng/mL correspond to deficits from alcohol of .05% and .08% respectively. This research stands out as an exceptional contribution because THC concentration cut-offs were calibrated against typical BAC cut-offs. With such data, the authors easily could conduct an additional study on the validity of any THC cut-offs for performance deficits. None of the studies below provide any concrete assessment of the validity of different cut-offs for impairment.

Ramaekers et al. (2006)—conducted a double-blind ***cross-over*** design of 20 subjects who smoked two different doses of cannabis (25 mg and 50 mg THC) and a placebo. Serum blood THC were paired with three measures taken at baseline and nine time points: distraction (a critical tracking test), a measure of impulse control (the stop signal test), and a measure of planning and problem solving (the Tower of London decision-making test). The authors concluded that serum concentration levels between 2 and 5 ng/mL could establish the lower and upper range for THC impairment. This study deserves attention because the findings have been influential in the field.

Although the study design was good, errors were made in the analysis that do not justify the overall conclusions. In particular, the authors used an improper statistical procedure that increases that likelihood of finding statistical significance and improperly interpreted the meaning of the relationship. The authors presented scatter plots of log transformed THC serum concentrations (not actual THC levels) and performance measures. The correlation coefficients from the regression analyses between log transformed THC and these three performance measures were poor (r-values ranged between .15 and .4, which means 2% and 16% of the variation in blood tests were related to the performance measures. These low r^2 values were in fact inflated by taking 11 different measures of the same individuals, a procedure that reduces the variability of the sample, leading to erroneously high correlations and violation of the assumption of independent measures required for the statistical test (see Chapter 3 of this book, "Understanding Statistics/correlations"). The proper statistical procedure for this kind of data is a mixed repeated measures ANOVA (Analysis of Variance). This study shows that blood THC levels are not an accurate predictor of the magnitude of performance deficits of these three measures. As well, the measures used may not translate into road safety concerns in the real world, a limitation that the authors do acknowledge.

The authors' overall conclusion is not justifiable that serum concentration levels between 2 and 5 ng/mL could establish the lower and upper range for THC impairment. The conclusion is not acceptable based on analytical shortcomings and unacceptable interpretations of the results from their study.

Khiabani 2006—In this study, 456 blood samples positive for THC only were identified from Norwegian drivers suspected of driving under the influence of drugs. The authors found a significant relationship between the proportions of subjects with judgements of impairment by police and higher THC positive groups. For example, they found that 38% of those with THC levels less than .7 ng/mL were judged as impaired compared to 57% for those with more than 10.2 ng/mL. Although this study does affirm a positive relationship between THC concentrations in blood and the likelihood of being assessed as impaired, this weak relationship found cannot be used to establish an accurate per se limit.

Schwope et al. (2012)—In this rarely cited study, 10 *heavy and chronic* cannabis smokers were administered a cannabis cigarette containing 54 mg of THC and followed up with performance and THC blood tests over six hours. Performance based on critical tracking and divided-attention tasks did *not* significantly change from baseline to 15 minutes after smoking. Blood THC concentrations decreased rapidly in the first hour after smoking, similar to finding reported by Hunault et al. (2014). The authors created a cannabis influence factor scale, based on THC, hydroxy-THC and carboxy-THC; however, they did not find this scale was useful for quantifying psychomotor impairment. This latter finding is not surprising, given that no meaningful performance deficits were found after cannabis administration.

Curran et al. (2002)—The effects of 15 mg and 7.5 mg *oral doses* of THC were compared with placebo capsules, in a cross-over design. Subjects were assessed at numerous time points, including pre-administration and one, two, four, six, eight, 24 and 48 hours afterward. Average blood plasma THC peaked at about 4.5 ng/mL at two hours and declined close to zero by eight hours. Significant performance deficits on two memory tasks, and subjective effects of being high, were found at two hours in the high-dose condition. Learning and working memory tasks were not related to doses. In terms of objective effects, significant drug effects were found on four tasks and ***non-significant*** effects were found on seven tasks. Significant subjective effects were found in the high-dose condition at eight hours, compared to the placebo. No effects were found on any measure after 24 or 48 hours.

Ménétrey et al. (2005)—This double-blind cross-over study with 10 subjects on the relationship between blood tests and performance is unique for several reasons. The authors had the following conditions: two with hemp milk at 16.5 and 45.7 mg THC (the latter dose is very high), one with synthetic THC (at 20 mg THC) and a placebo condition. The authors took measures of both THC and hydroxy-THC, and incorporated an outcome of willingness to drive under different conditions. As mentioned earlier, THC concentrations were very low; around 8.4 ng/mL at peak concentration and declined very slowly thereafter for the 45.7 mg THC dose. Similar to studies of smoking cannabis, extremely large variations between subjects in THC and cannabinoid levels (hydroxy-THC and carboxy-THC) were found at the same time points after smoking. For example, the range in THC levels at 2.5 hours was from 1.6–9.0 ng/mL in the high-dose condition, and the ranges for hydroxy-THC at all time points were also extreme.

The relationship between THC levels in blood (for the 45.7 mg dose) and subjective reports of feeling high are shown in **Figure 5**. The subjects also gave subjective reports of how high they felt at different time points. I have converted the ratings from a 10-point scale to a 100-point scale to allow rough comparisons with the Hunault et al (2014) study. Average subjective highs of at least 80/100 were found between one and four hours. The following conclusions can be drawn when comparing the results of this study with the one by Hunault et al (2014).

The ratios of THC concentration amounts administered appeared to correspond more closely to THC levels in blood than with smoked cannabis, possibly due to titration while smoking. Major decrements in performance were found for the strongest concoction. Subjective ratings of being high peaked at just over 2.5 hours after use and lasted up to 10 hours. Most participants further indicated that they would not drive a friend to a party after 10 hours in the high-dose condition. Also of note, two subjects were withdrawn from the study due to adverse side effects of anxiety or vomiting. This study raises several issues regarding cannabis and driving. It appears that ingesting cannabis may produce more extreme performance deficits than smoking and over a longer period than smoking, despite the fact that subjects may have lower THC concentration levels than would occur from smoking. Cannabis smokers can titrate the amount inhaled, commonly observed among smokers (Hartman et al., 2015). The Ménétrey et al. (2005) study highlights the possibility that hydroxy-THC can cause performance deficits.

Figure 5: *The relationship between THC concentrations in whole blood and subjective feelings of being high for* ***ingested cannabis*** *of 45.7 mg THC*

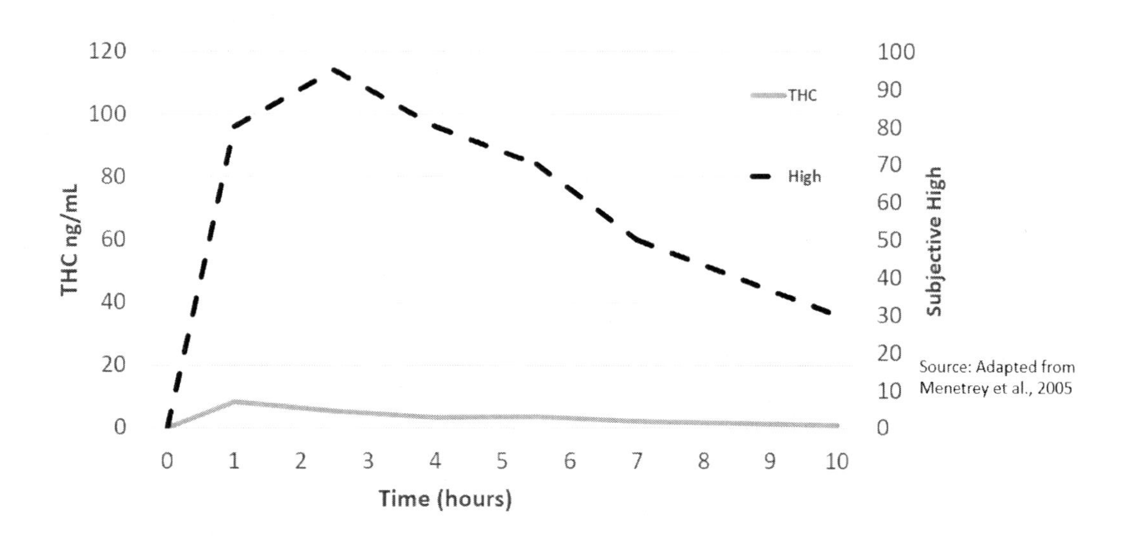

Interestingly, concentration levels of hydroxy-THC were also taken. Recall from Chapter 4 that the liver converts THC into hydroxy-THC, which has psychoactive properties, possibly more potent than THC itself. Hydroxy-THC levels exceeded THC levels at every time point, peaking at 12.8 ng/mL at 2.5 hours, the same time point of the greatest subjective high. At 10 hours in the high-dose condition, average THC levels were only .9 ng/mL, whereas hydroxy-THC levels were 2.5 ng/mL. THC and hydroxy-THC levels combined provided the best estimate of impairment. While this conclusion is of interest and implicates hydroxy-THC as an important psychoactive metabolite for impairment, more research is needed on the role of hydroxy-THC in impairment.

Lenne et al. (2010)—This double-blind ***counterbalanced design*** study aimed to assess the performance effects of cannabis under placebo, low and high-dose conditions. Driving simulations in the cannabis dosages conditions were related to decreased speed and longer reaction times compared to the placebo condition, and blood tests confirmed higher average THC concentration in the low (7.4 ng/mL) and high-dose (12.01 ng/mL) conditions. Like most other studies, however, performance thresholds that might constitute impairment were not established and direct measures of paired blood and performance measures were not reported.

Interestingly, average blood THC levels after smoking high, medium and low doses of THC do not appear to vary in the same proportions as THC smoked, although high-potency THC does produce higher THC blood levels. This observation is also consistent with a study by Hunault et al. (2014). Peak average THC serum levels in the high THC condition (69 mg) cigarette was about 180 ng/mL, whereas in the low level (29 mg) cigarette, the peak was about 160 ng/mL. Another way of stating this is that the dosage ratio was about 2.4/1 (i.e. 69/29), whereas the peak blood THC ratio was about 1.2/1 (180/160). Titration may be an explanation for these divergent ratios. When eating cannabis, THC levels in blood tend to reach a much lower peak level at about one hour after ingestion.

THC in oral fluids (OF)

Fierro et al. (2014)—The authors examine the relationship between oral fluid (OF) tests and observations of impairment by police officers. Observations were based on a roadside survey of over 3,000 Spanish drivers. The police officers received 20 hours of training, which included identification of 31 signs of being under the influence of cannabis. OF concentration levels based on confirmatory test of THC were categorized into five ordered ranges. This is the only study I am aware of that directly compares OF concentration categories with observations of impairment.

I have re-analysed the data from the Fierro et al. (2014) study to illustrate the type of analyses that are helpful for deriving evidence-based per se limits to detect cannabis impairment using epidemiological methods to assess the validity of OF cut-offs of 25 and 100 ng/mL THC against the standard of observations by trained police. **Table 7** below shows the relationship between OF tests at a cut-off of 25 ng/mL. This analysis shows that at two cut-offs, accuracy is good, sensitivity is poor and specificity is good—largely influenced by the very high proportion of subjects with no THC present. Positive predictive value, the likelihood that someone who tests positive will show any sign of impairment, is poor at 19.6% with a 25 ng/mL cut-off and 22% at a 100 ng/mL cut-off. In the above examples, using expert detection of **any** signs of impairment and an OF cut-off levels of 25 ng/mL and 100 ng/mL, both Kappa values are below .4, which represents poor agreement (Gordis, 2014; p. 110). Overall, the relationship is unacceptable. The improvement in accuracy at a cut-off of 100 ng/mL versus 25 ng/mL is minimal. If a valid cut-off for oral fluids to assess impairment exists, the concentration level is likely substantially greater than 100 ng/mL. No biomarker cut-off from cannabis has been validated for impairment. The definition of impairment by cannabis by behavioural criteria has been elusive and no gold standard similar to SFST for alcohol exists.

Table 7: *The relationship between any impairment signs and oral fluid tests for THC at 2 cut-off levels**

25 ng/mL cut-off	Any sign	No sign	Total	100 ng/mL cut-off	Any sign	No sign	Total
>25 ng	27 (a)	111 (b)	138	**>100**	20 (e)	69 (f)	89
OF <= 25 ng	26 (c)	2235 (d)	2261	**<= 100 ng**	33(g)	2277(h)	2310
Total	53	2346	2399	**Total**	53	2346	2399

**Detailed calculations for 25 ng/mL:*

(a) = (49*14.3%) + (89*22.5%) = 27
(b) = (89+49) – 27 = 111
(c) = (2146*1%) + (34*2.9%) + (81*4.9%) = 26
(d) = (2146+34+81) – 26 = 2235

(e) = (89*22.5%) = 20
(f) = (89 – 20) = 69
(g) = (2146*1%) + (34*2.9%) + (81*4.9%) + (49*14.3%) = 33
(h) = (2146+34+81+49) – 33 = 2277

Example 1 at 25 ng/mL—Accuracy = 94%, Sensitivity = 27/53 = 51%, False negatives = 26/53 = 49%; Specificity = 2235/2346 = 95%, False positives = 111/2346 = 5%; Positive predictive value = 27/138 = 19.6%; Kappa = .256 (95% CI, 0.175 to 0.343)

Example 2 at 100 ng/mL—Accuracy = 96%, Sensitivity = 20/53 = 38%, False negatives = 33/53 = 62%; Specificity = 2277/2346 = 97%; False positives = 69/2346 = 3%; Positive predictive value = 20/89 = 22%, Kappa = .261(95% CI, 0.163 to 0.359).

Ramaekers et al. (2006)—In this study, both serum blood tests and oral fluid test were taken, along with three measures of performance at 10 time points, previously described above. As noted previously, the conclusion by the authors that 2–5 ng/mL establish a per se range is not justified by the methods. No recommendations were provided with respect to impairment by oral fluids; however, findings from this study should be given little weight based on the flaws previously described.

Carboxy-THC in urine

I am not aware of any laboratory studies on the relationship between urine test results and performance deficits. Given our current knowledge of the detection periods for components and metabolites in urine, it is unlikely that such research would yield worthwhile findings. Compounds are detectable in urine two to six hours after the use of cannabis and other drugs (Swotinsky, 2015; p. 216). After smoking cannabis, the most intense high occurs within the first hour or two, precisely the period of most relevance to detect impairment.

CONCLUSION

Overall, there is scientific consensus that acute use of both alcohol and cannabis causes performance deficits that warrant prohibiting driving while impaired. Most laboratory studies find that performance deficits become substantial at a BAC level of .05%. One can conclude from existing research that after smoking cannabis, subjects remain high for at least the first hour when THC levels spike. From these observations, it appears reasonable that a THC whole-blood cut-off between 10–20 ng/mL might constitute cannabis impairment when cannabis is smoked. However, when cannabis is ingested, much lower THC levels plus an additional metabolite, hydroxy-THC, appear related to impairment. Existing research presented in this chapter points toward a much higher THC cut-off in blood after smoking than currently exists in most countries with this type of legislation.

One major limitation of all the studies found on performance deficits and THC blood levels is that none of the studies conducted an assessment on the validity of a particular cut-off, a standard epidemiological approach outlined in Chapter 2 of this book. Hartman et al. (2015a) was the only study that attempted to calibrate their findings into meaningful criteria by calculating blood THC levels that are comparable to common thresholds from alcohol. They found 8.2 ng/mL and 13.1 ng/mL correspond to deficits from alcohol of .05% and .08% respectively, which is still higher than cut-offs in most countries. Data from this study and others could be re-analysed to assess validity and the likelihood that drivers at these THC levels are truly impaired. Such an assessment is needed to accommodate individual differences.

Although the research examined in this chapter does show meaningful performance deficits from the acute effects of cannabis that could be considered impairment, the research on a cut-off level from biological tests to detect impairment is scant. Studies on the relationship between blood tests for THC and impairment are more common than those with oral fluid testing but neither type of biological test has been shown to be sufficiently valid based on traditional epidemiological methods. None of the studies have used traditional epidemiological approaches to assess the validity and reliability of a biological cut-off level for determining impairment. Biological tests need to be assessed against a more accurate external standard of impairment in order to properly assess validity and reliability. Unfortunately, there is a gap between these recommendations and the experimental literature on performance deficits, as it is not always evident how the magnitude of an effect from laboratory studies translates into elevated crash risk on the road.

CHAPTER 6

Long-term effects of alcohol and cannabis

HANGOVER, WITHDRAWAL & LONG-TERM EFFECTS

In this chapter, I describe long-term effects, including withdrawal, hangover and lasting cognitive deficits, of alcohol and cannabis. Legislation does not exist for the use of biological tests to assess the hangover effects of alcohol for drivers. Much of the existing and proposed legislations for cannabis worldwide set thresholds of 5 ng/mL or lower for THC in blood, which could identify substantial proportions of drivers who are not actually impaired. One rationale for low thresholds is that long-term deficits are a concern for safety on our roads. In fact, there have been studies of long-term deficits from using cannabis. I compare the scientific evidence of deficits from alcohol and cannabis in terms of long-term performance deficits in the following pages.

Alcohol

Characteristics of an *alcohol hangover* include a constellation of unpleasant physical and mental symptoms, especially a headache and irritability, which occurs between 8 to 16 hours after heavy drinking (Kim et al., 2003). There is some evidence of specific cognitive deficits in attention and memory during an alcohol hangover, but this may only be true for severe hangovers (Ames et al., 1997; Howland et al., 2010; Ling et al., 2010). Kim et al. (2003) reported a significant decrease in visual, memory and intellectual processes during a hangover state when compared to baseline performance, whereas Finnigan et al. (1998) found no evidence of impaired psychomotor performance. In a review of the scientific literature, Wiese et al. (2001) found that although hangovers are experienced by those with alcohol dependence, light-to moderate drinkers incur most of the costs due to hangovers. A review of the experimental literature on hangover and performance found 27 peer-reviewed studies; only eight were judged methodologically rigorous enough to be included in the review (Stephens et al., 2008). Only two of these eight studies found performance deficits related to alcohol hangovers. However, there is evidence of measurable decrements in performance in some recent studies (McKinney et al., 2012; Verster et al., 2014). Overall, research evidence indicates that alcohol hangovers can be associated with performance deficits, but higher-quality research is needed (Verster et al., 2010).

Withdrawal effects for alcohol are usually moderate but can be severe with heavy long-term usage. Sleep disturbances and mild anxiety are common, severe cases of withdrawal result in delirium, seizures and convulsions.

In terms of *long-term effects,* evidence suggests that some long-term cognitive deficits from alcoholism can linger for up to a year (Stavro et al., 2013). Cognitive functioning and brain damage can result from extended periods of heavy alcohol use. Extreme usage can result in Korsakoff syndrome—irreversible brain damage characterized by memory loss (National Institute of Alcohol Abuse and Alcoholism, 2001). Milder deficits related to complex visual-motor functions, memory and problem solving are also possible with heavy consumption (Coambs & McAndrews, 1994).

Cannabis

Characteristics of studies showing a hangover effect

Hangover effects, commonly referred to as *carry-over* or *residual effects* from single acute exposure, have been studied for cannabis. Key findings and methods are depicted in **Table 8** and summarized briefly here.

Yesavage et al. (1985) was the first randomized study to report carry-over effects 24 hours after smoking cannabis. It was described as a preliminary study (in preparation for a larger study) and as such, their findings should be given little weight. A subsequent study of airplane pilots was conducted (Leirer et al., 1991) but had severe flaws, including lack of randomization, no blinding of conditions and selective dropping of measures after data collection. Also, a preliminary study by Heishman et al. (1990) should be given little weight, as it only had three subjects and no statistical tests of significance. These three studies, all with major limitations, reported long-term performance deficits from cannabis. None of the studies could adequately explain the meaning of their findings in the real world.

Table 8: *Studies of hangover effects from cannabis*

First author, year	Randomized or counter-balanced?	Double blind	Limitations	# subjects	Deficits, time over 4 hours
Yesavage et al., 1985	Yes	Yes	Preliminary	10	Yes, 24 hr
Chait et al., 1985	Yes	Subjects blinded		13	Yes, 9 hr
Leirer et al., 1989	Yes	Yes	Measures dropped	18	No
Heishman et al., 1990	No	No	Preliminary, no statistics	3	Yes
Chait, 1990	Yes	Yes		12	No
Leirer et al., 1991	No	No	Measures dropped	9	Yes, 24 hr
Fant et al., 1998	Yes	Yes		10	No
Curran et al., 2002	Yes	Yes		15	None at 24 or 48 hr
Ronen et al., 2008	Yes	Yes		14	No

Characteristics of studies not showing a hangover effect

Several other studies with stronger methodologies reported no long-term deficits after smoking cannabis. In a follow-up study with double-blind and randomized conditions, no carry-over effects in the cannabis condition were found at 24 hours. In a randomized study by Chait (1990), there was no evidence of cognitive deficits the morning after smoking cannabis. Later, Fant et al. (1998) conducted a randomized double-blind study and concluded, "No effects were evident the day following administration, indicating that the residual effects of smoking a single marijuana cigarette are minimal" (p. 777). Curran et al. (2002) administered oral doses of THC (15 and 7.5 mg) in a well-controlled study and found no effects on about 20 different measures at 24 and 48 hours. Ronen et al. (2008) conducted a double-blind counterbalanced study and reported, "No THC-related effects were measurable 24 hours after smoking the high (17 mg) level of THC" (p. 933).

The aforementioned studies are of sound methodological quality and show deficits after using cannabis are unlikely at 24 hours. None of the higher-quality studies (Chait, 1985; Leirer et al., 1989; Chait, 1990; Fant et al., 1998; Curran et al., 2002; Ronen et al.,

2008) found deficits after 24 hours, although one study (Chait et al., 1985) found measurable deficits at nine hours. The remaining studies that found effects either did not have a randomized design (i.e. Heishman et al., 1990; Leirer et al., 1991), were described as preliminary (Yesavage et al., 1985; Heishman et al., 1990), or had no statistical tests of significance (Heishman et al., 1990).

Preponderance of the evidence

Despite the preponderance of the evidence that does not point to meaningful deficits after 24 hours, the Leirer et al. (1991) study, which says cannabis causes 24-hour deficits, is frequently cited. This study has multiple methodological issues, such as no random assignment, failure to conduct a pre-test of the methods, and retention of unreliable measures briefly mentioned in an earlier review (Macdonald et al., 2010). In terms of assignment to groups, the authors indicate subjects were assigned to either the marijuana or placebo condition, always referring to the marijuana condition first; and they used a 30 day delay between sessions to minimize practice effects. If this was the true order of conditions, then the marijuana condition is *confounded* with practice effects (i.e. subjects do better on tasks with more experience). Instead of randomization, the gold standard of experimental studies, the authors reported that practice effects were addressed by keeping the sessions 30 days apart. In addition, two measures (far target detection and engine malfunctions) were dropped from the analysis after data collection was complete because they were not reliably recorded. The proper procedure for a study is to decide on valid measures at the onset of the study and to keep them throughout the study. No blinding of researchers was conducted, which opens up the possibility of researcher bias. In the earlier study (Leirer et al., 1989), the authors did specify both randomization and blinding, and found no hangover effects. When these methodological criteria were omitted from the study, deficits at 24 hours were reported.

Interestingly, the data for near targets was kept, even though exactly the same procedures were used to collect this data as far-target detection. Furthermore, near-target avoidance data appears unreliable as follows. Performance scores in this study were expressed as decrements so that larger values represented greater decrements, and the marijuana condition was subtracted from the placebo condition. At .25 hours, four hours, eight hours, 24 hours and 48 hours the corresponding difference scores were 37.32, -2.46, 8.88, 49.94, and -27.63. At 24 hours, performance was worse than the most likely peak of impairment time of 15 minutes for the marijuana group. At four hours, those in the marijuana condition did better than those in the placebo condition. These data are unreliable because there is no trend in scores over time, yet these scores were included in the analyses and helped show statistical significance at 24 hours.

Despite clear methodological limitations of this study, its findings have been uncritically disseminated in peer-reviewed journals. Several authors simply ignored the findings from other studies with better methods and no long-term deficits, opting to cite from the methodologically weaker Leirer et al. (1991) (Couper and Logan, 2004; Solowji et al., 1995; Ashton, 2001; O'Kane et al., 2002). O'Kane et al. (2002) went even further by describing the Leirer et al. (1991) study as a ***cross-over trial***, despite the fact the authors of the original study never provided such a description. In two review articles, Ashton (1999 and 2001) uncritically accepted the findings of this study by replicating data. The aforementioned examples illustrate how myths can be propagated and maintained over time. The proper way to assess a research issue is to weigh the balance of the research findings in relation to the methods used.

Leirer et al. (1989) is a good example of a study with issues to do with translating findings from artificial laboratory studies to the real world. This study examined the relationships between age, marijuana doses and simulated aircraft performance with conditions in randomized order. An interesting finding of this study was that young subjects in the high-dose (20 mg) marijuana condition performed better than older subjects in the placebo condition at every time point (see Leirer et al., 1989; p. 1149). This conclusion is based on visual inspection of the means between groups. No statistical tests of significance were conducted to assess the likelihood this relationship was due to chance, as this type of conclusion was likely of little interest to the authors. The authors did report significant ($p<.05$) performance decrements associated with cannabis at one hour and four hours compared to the placebo condition. One issue in this study is: how meaningful can these deficits be in the real world when younger subjects in the high-dose marijuana condition did better than older subjects in the placebo condition at every time point? You would think from these results, young men would be safer in the real world than older men; however, studies in real-world conditions have found young men have greater crash risk and job accident risk than older men (Insurance Institute for Highway Safety, 2013; Swedler et al., 2012). The findings from this study is a complete contradiction to what occurs in the real world with respect to age and accident risk and therefore have little relevance.

Some researchers have reported withdrawal effects from cannabis, but the scientific evidence that withdrawal symptoms constitute meaningful impairment is weak. For example, Budney et al. (2001) describe withdrawal from cannabis as similar to withdrawal from nicotine. Substantial individual differences exist in terms of the nature and severity of withdrawal symptoms. Smith (2002) conducted a literature review and concluded that, "there is not a strong evidence base for the drawing of any conclusions as to the existence of a cannabis withdrawal syndrome in human users" (Smith, 2002; p. 621). In a more recent study of 469 adult cannabis smokers who made a quit attempt, 42.4% reported withdrawal symptoms (Levin et al., 2010).

Long-term cognitive deficits from long-term heavy cannabis use are possible based on studies of long-term abstinent cannabis users versus controls. For example, Crean et al. (2012) conclude there are long-term decision-making and risk-taking deficits associated with long-term use based on a single study. Other studies found stopping long-term heavy use of cannabis may result in subtle but not clinically disabling cognitive deficits (Pope et al., 2002; Solowij, 1995). Gruber et al. (2003) found that heavy users of cannabis reported lower levels of educational attainment and income than controls. These studies have major weaknesses in research design and the magnitude of deficits found do not appear substantial. Much of the research compares cannabis users who quit with non-using controls. This is methodologically unsound, as cannabis users may differ from non-cannabis users in many respects which may be precursors to using cannabis, such as impulsiveness and mental health issues. Cannabis use may be an outcome of other individual traits. It is not clear that such deficits would be extreme enough to pose a meaningful safety risk while operating potentially dangerous equipment or driving a car. Not surprisingly, Broyd et al. (2016) concluded that evidence that long-term cognitive deficits persist after abstinence is weak or mixed. If cognitive deficits do exist, the magnitude of these effects is likely minor and not of clinical importance.

CONCLUSION

Preponderance of findings from studies on hangover effects indicate meaningful performance deficits from alcohol hangovers but not from cannabis. Withdrawal symptoms creating performance deficits from dependence can be mild to severe for alcohol and from mild to moderate for cannabis. Major cognitive deficits can result from long-term heavy use of alcohol but the research evidence for cognitive deficits from cannabis is weak.

CHAPTER 7

Observational studies of alcohol and cannabis risk in crashes

INTRODUCTION

An observational study, also called a field study, is a type of research study that is conducted in real-world environments. These studies and methodological issues with them are the focus in this chapter. They differ from laboratory studies in that researchers do not administer substances or assign subjects to different conditions but rather cannabis use is detected in natural environments. Case-control and ***culpability studies*** are types of observational studies that seek to determine whether substance use is over-represented among crash-involved cases. In the case-control method, crash-involved drivers are compared to non-crash-involved drivers in terms of detection of substances at various thresholds. In the many studies the cases and controls are matched for age, sex and location of crash, and the presence of other substances. Another related approach used for fatal crashes is responsibility or culpability analysis where drug substance involvement is compared between those responsible and not responsible for crashes.

Findings from evaluation studies, treated as a class of observational studies, are also examined in this chapter. This kind of study involves real-world interventions, such as per se laws, intended to reduce crashes and typically, rates of crashes, ideally involving substance use, are compared between two time periods.

EPIDEMIOLOGICAL GUIDELINES FOR VALID FINDINGS

As an epidemiologist, there are several criteria used in order to interpret the meaning of findings from a study. The main issues of importance are described below.

(1) Bias

Bias is a systematic error in the design or conduct of study that leads to an erroneous association between the drug use (i.e. exposure) and crashes (i.e. disease). Findings from studies with major sources of bias should be given little weight in drawing conclusions. Several types of bias can occur in observational studies.

A common source of bias in many case-control studies examining the relationship between acute effects of drugs and crashes is ***selection bias***. This type of bias, due to non-respondents, is created when drug users in the control group are less likely to participate in a study than the cases, a limitation noted by Moskowitz (1985). This limitation will also have the effect of creating inflated ***odds ratios (OR)***. It is particularly an issue where consent to participate in the study is required from the control subjects but not the cases. This situation typically occurs when fatally injured drivers, where consent is not required, are compared to living control subjects, where consent is required. In one study, of where randomly selected control subjects were asked to provide a urine sample, only 49.6% consented (Dussault et al., 2002). One could reasonably expect that a high proportion of drug users were among those who refused to participate. This means that drug use is lower than expected in the control group due to the design of the study. Drug users can be over-represented in the cases compared to the controls, causing an inflated but erroneous OR.

Another common type of bias occurs when the cases are not representative of the larger population of drivers in crashes. This occurs when fatally injured drivers who are more likely than others to test positive for substances are selected for biological drug tests. This is very common occurrence. In North America, and likely many other countries worldwide, biological tests of fatally injured drivers are not mandatory; only a subset is selected for drug tests. Bias occurs because drivers who were more likely to test positive are selected by coroners for drug tests (i.e. young men). At least a couple of studies show that tests were much more likely to be taken for young drivers and men in fatalities (Brault et al., 2004; Li et al., 2013). Since both men and 16–24-year-olds are more likely to use cannabis and alcohol than others (Statistics Canada, 2012), this bias also would lead to a greater proportion of drug users (i.e. an unrepresentative sample) in the driver fatality group, and have the ultimate effect of artificially and erroneously inflating the ORs.

Culpability studies are advantageous over case-control studies of drug users because they are less likely to be biased. I suspect it is for these reasons that, in a recent meta-analysis by Asbridge et al. (2012), all of the studies identified as "high quality" were culpability studies. Culpability studies are advantageous in that typically both cases and controls are selected using the same methodology, which eliminates potential biases. A drawback is that these studies include fatally injured drivers and the findings may not be applicable to drivers involved in non-fatal crashes.

I only know of two case-control studies of non-fatal crashes that were successfully able to eliminate the likelihood of selection bias (Marquet et al, 1998; Brubacher et al., 2016). In both these studies, consent to draw and test biological specimens was not required by the ethical review boards. Marquet et al. (1998) obtained urine samples from hospitalized drivers in crashes and controls. Brubacher et al. (2016) conducted a ***culpability study*** based on excess blood taken from hospitalized subjects who required these samples for their treatment. The authors implemented anonymization procedures so that the identity of subjects could not be traced to the drug test results.

It should be noted that bias has been practically eliminated for case-control studies of alcohol risk in crashes. The main reason for this difference is that alcohol impairment is easy to detect and it is a criminal offence to drive while impaired. In a recent ***case-control study*** by Blomberg et al. (2005), 97.9% of control subjects provided a breath sample, a percentage much greater than studies on drug use (for example in the Brault et al. (2004) study, only 49.6% provided a urine sample). Typically, control drivers who provide a breath sample over the legal limit are not charged and given a ride home because they agree to be part of a research study. Hence, there is an incentive for them to participate, especially considering that all these studies require police involvement to collect data on the road. As there is no per se limit for charging drivers under the influence of cannabis and other drugs, this means that there are essentially no potential benefits for participating if the drivers use cannabis. Thus, those who use cannabis may be less likely to participate.

(2) Control for confounding

A confounder is a variable that is causally related to a condition or serves as a proxy measure for unknown causes; it is associated with the exposure under study, but not a consequence of exposure (see Kelsey et al., 1996). In research on crashes and substance use, three major variables are viewed as confounders: concurrent alcohol consumption, age and sex. These variables are associated with both cannabis use and crashes. For example, males, younger people and those who drink and drive are more likely than others to be responsible for crashes and to drive after using cannabis. The challenge in research is to separate out the effects of cannabis from these confounders. This is typically addressed by a matching procedure in the design of a study (e.g. ensuring the same proportions of males in the cases and controls) or through statistical control of these variables in ***multivariate*** analytic procedures. When authors indicate they conducted multivariate analyses or adjusted tests, this means they have adjusted their findings to account for selected ***confounding variables***. When evaluating whether findings are a good reflection of reality, epidemiologists pay particular attention to potentially confounding variables and how they might influence the findings of the study if not properly addressed. When studies report crude or unadjusted ORs, this means that no consideration for any potentially confounding variables has been taken into account for the analyses. This kind of analysis should be treated cautiously. If confounding variables are not taken into consideration when reporting findings, typically the ORs will be erroneously inflated.

(3) Measurement error

A type of measurement error that is extremely common in cannabis crash research is drawing conclusions about impairment of drivers based on drug tests that are invalid for this purpose. This shortcoming is compounded when different methods to measure drug use are treated as equivalent, despite ample scientific evidence that the measures are fundamentally different. Drug use can be assessed many ways: self-reported impairment, self-reported past use, dependency, biological specimens (urine tests, blood tests, oral fluid tests), inactive metabolites (e.g. carboxy-THC) versus THC, immunoassay vs. GC/MS, and different cut-offs. Several studies have used mixed measures to examine crash risk by taking different types of samples from the cases than from the controls—for example, blood samples in experimental cases and oral fluid or urine samples in the controls. Since science is not advanced enough to conclude equivalency between these measures, such studies should be given little to no weight when drawing conclusions.

(4) Strength of association

Another principle of epidemiology is that a strong association provides more compelling evidence of ***causation*** than a weak association. Different statistical tests are used to measure the strength of associations (technically called ***effect sizes***) based on how the variables are measured and on the study design. The most common tests are correlations (r) for interval or ratio level data, odds ratios (OR) for dichotomous variables in cross-sectional designs, and relative risk (RR) for dichotomous variables, appropriate in ***cohort studies***. For OR and RR values, the range of values is from 0 to infinity, with values from 0 to 1 representing a protective effect and over 1 representing a risk factor. A value of exactly 1 indicates no relationship. Some authors treat ORs and RRs as synonymous, but they are not. ORs are calculated by comparing the *ratio* of those testing positive to negative in the performance deficit group versus the same calculations in the control group. RRs are calculated by comparing the *proportions* of those testing positive in accidents (or responsible) for all positives versus those testing negative in accidents (or responsible) for all to those testing negative. Ideally, factors such as sex, age and risk-taking propensity (confounders) should be ruled out as possible explanations for the results found, typically achieved through ***multivariate statistical methods***.

The likelihood that these tests of association could be explained by chance factors is assessed through statistical procedures, typically with confidence intervals (CI) or ***probability values***. Studies with larger sample sizes can detect smaller differences between groups. If the confidence interval does not overlap with one or the p-value is less than .05, the association is significant. A confidence interval can be interpreted such that if the study was conducted 20 times with the exact same methods, 19 of the estimates would fall within the range. A significant probability test is a necessary condition to conclude that there is a statistical association between two variables. However, examination of the strength of the association (i.e. effect size) is also needed to conclude a relationship is meaningful. Such data can demonstrate associations, but in themselves do not show causation. Strong conclusions of causality can be drawn when findings from both experimental laboratory-style studies and observational studies converge to show the same statistical relationships. The likelihood that there is an alternative explanation for the observed effects that can be explained by the confounding variable in observational studies is based partially on the magnitude ORs or RRs. I define a strong effect as an OR or RR of over 5, a moderate effect of 4 to 5, and a weak effect of between 1 and 3 (Macdonald, 2015; p. 103). Protective effects (0 to 1) can be assessed in the same manner as reciprocals. For example, the reciprocal of an OR of .2 equals 1/.2 or 5, which is a strong effect size. A summary of the strength of relationships for different ORs is provided in **Table 9**.

Table 9: *Interpretation of effect sizes from crude Odds Ratios (OR)*

Strength of relationship	Risk factor	Protective factor
Weak	Between 1–3	Between .33–1
Moderate	Between 3–5	Between .20–.33
Strong	Above 5	Under .20

Weak ORs are extremely common in research and often are interrelated with other variables that are causal. This point is illustrated in one of my studies, where I calculated ORs for a number of variables in relation to job injuries (Macdonald, 1995; Table 1, p. 706). I found 28 variables were significantly ($p < .05$) related to job injuries, with corresponding odds ratios of 1.8 or greater. These variables included employees who feel they do too much shift work, feel their work is boring, are in conflict with others or smoke cigarettes. Importantly, crude ORs of these magnitudes can easily disappear when controlling for confounding variables and should not be interpreted as causal (Remington et al., 2010; p. 51). By contrast, risk factors that represent strong associations and compelling evidence of causation include Blood Alcohol Content (BAC) and increased likelihood of traffic crashes (where the RRs rise exponentially to over 25 with increasing BACs), and smoking cigarettes and lung cancer (OR of 15–30, depending on dosage) (US Department of Health and Human Services, 2014).

(5) The difference between causation and correlation

An important objective of epidemiological studies is to determine whether relationships between variables (e.g., drug use and accidents) are causal. Causality means that a change in one variable will alter another variable. In epidemiology we are studying populations were causal factors differ among people. Therefore, we examine general caustation. There are three requirements in order to demonstrate general causation. There are three requirements in order to demonstrate causality in epidemiological studies: the suspected cause must precede the suspected effect in time, a statistical relationship between two variables of interest must exist, and the observed empirical relationship cannot be explained by a third variable. This third requirement is the most challenging to test and is related to the issue of confounding, described above. Therefore, while an association may exist between two variables, this relationship is not necessarily causal. This is of particular importance when assessing whether substance use causes crashes. Some studies have found a relationship may exist between drug use and crashes; however, little or no efforts have been made in most studies to assess whether the observed relationships are causal. In fact, the relationship between drug users and accidents/injuries might be better explained by other variables. Generally, given proper methods, the stronger the relationship, the more likely it is causal. Crude ORs with weak effect sizes, as shown in Table 9, should not be used to imply causal relationships.

A question of great importance and one that is addressed extensively in this book is which criteria were used to establish the standards for alcohol. Clearly, there will be individual differences in terms of performance deficit for individuals at or above this level, but the question of importance is the establishment of a critical threshold whereby the *majority* of people will experience meaningful performance deficits.

WHO IS LIKELY TO DRIVE IMPAIRED—AND HOW DO THEY BEHAVE?

Two issues of importance for understanding the relationship between impaired driving and crashes, and appropriate interventions is who is likely to drive impaired and how do they behave. People who use substances have other characteristics that can increase their likelihood of crashes, such as being male or younger. A challenging issue in research is separating out the importance of different risk factors from substance use in explaining the causes of crashes.

Alcohol

With respect to who is likely to drink and drive, studies have consistently shown that those who are young, male or heavy drinkers are more likely to drive while impaired by alcohol and to be involved in alcohol-related crashes (Macdonald, 2003; Macdonald & Pederson, 1988). Interestingly, males and younger drivers are the two groups with an elevated likelihood of crashes without alcohol. Also, risk taking and some personality characteristics, such as aggression, are related to an increased likelihood of crashes and likelihood of using. These intertwined relationships make it challenging to separate whether alcohol use or some other confounders contributed to the crash.

The next question of importance is how people behave when they drive while impaired. It was noted several decades ago by Gusfield (1979) that no studies have examined how people do drive after drinking. Empirical evidence shows that some drinking drivers are able to evade detection better than others, and that those with multiple ***Drivng while impaired (DWI)*** arrests tend to have more risky driving styles while impaired than those without arrests (Macdonald & Pederson, 1988). In the 1990s, it became clear that hard-core drinking drivers contributed an excessive proportion of drinking-driver crashes (Simpson & Beirness, 2004). They are characterized by alcohol problems, antisocial behaviours and worse driving records. Robbe and O'Hanlon (1993) and some other authors have suggested that drinkers underestimate their impairment and do not adjust their driving accordingly. Some drivers do appear more likely to be apprehended than others. From research of alcoholics (now called alcohol use disorders; Nordqvist, 2018) in treatment, I estimated the sample drove while impaired by alcohol about 8.6 days per month, and that the likelihood of being apprehended by the police was about 1 in 1168 drinking-and-driving events (Macdonald & Pederson, 1988). Those with multiple Driving While Impaired arrests reported more risky styles of drinking and driving than those with no arrests and have more collisions. This later finding does suggest the possibility that a subset of heavy drinkers can compensate for their impairment to reduce their likelihood of being apprehended.

Cannabis

Research similarly indicates that those younger, heavier users, and males may be over-represented in cannabis crashes. Again, younger people and males are more likely than older people or females to both get in crashes and to use cannabis. These are two potential confounders in traffic accident research that need to be properly addressed in research studies. The importance of youth and sex needs to be separated from the role of cannabis in crashes.

Additional evidence indicates that those who experience greater negative effects from cannabis are less likely to drive under the influence than those who experience milder effects (Macdonald et al., 2008). In this study, among 149 clients in treatment for cannabis abuse who reported driving while under the influence, 34.2% indicated they drove more carefully or cautiously (Macdonald et al., 2008). Smiley (1999) also reported for those that do drive under the influence of cannabis may drive more cautiously than when sober (and Robbe and O'Hanlon (1993) suggested cannabis users could compensate for their condition). One laboratory study found decreases in average speed and a longer average distance when following cars in the cannabis conditions (Hartman et al., 2015b). Although compensation to some degree is possible, not all performance deficits can be compensated.

ALCOHOL

Borkenstein et al. (1964) conducted the Grand Rapid study on the relationship between BAC and crash risk using a case-control methodology. This landmark study found an exponential increase in crash risk with increasing BACs above .04% alcohol.

Virtually every large case-control study since Borkenstein's study has shown a very strong relation between blood alcohol content (BAC) levels and the likelihood of collisions. Relative risk functions from several studies are depicted in **Figure 6** (adapted from Donelson and Beirness, 1985, and Blomberg et al., 2009). Epidemiological studies have found a strong dose-response relationship between greater doses of alcohol and crash risk. Consistent with these epidemiological studies are experimental studies that clearly show performance deficits with larger doses of ethanol.

Figure 6: *Observational studies on the relative risk of crashes at different BAC levels*

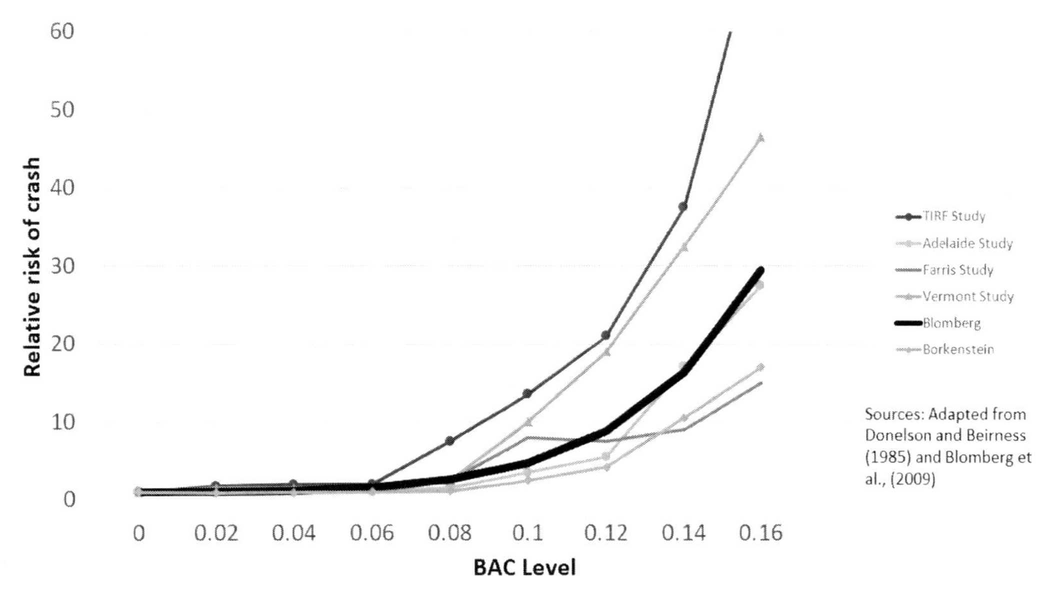

Analytic epidemiological studies on alcohol impairment and collisions are persuasive that alcohol is a cause of collisions (see Macdonald (1997) for a summary of this evidence). Virtually every large case-control study conducted has shown a strong relation between blood alcohol content (BAC) levels and the likelihood of collision. An exponential increase in collision risk is observed in all studies when BACs exceed .08% alcohol. Most recently, a high-quality, large scale study of BAC levels among drivers in 4,919 crashes and matched controls was conducted in two US cities (Blomberg et al., 2009). The modelled risk curve from the Blomberg study presented in Figure 6 is based on a final model, controlling for 33 covariates and ***imputation*** procedures. The final risk model from Blomberg et al. is likely overstate the risk of alcohol in crashes based on the imputation procedures used, explained in more detail in the next section. Given this limitation, the study still represents the most comprehensive case-control study ever conducted and therefore the risk curve from this study is bolded in Figure 6. Consistency of findings with different methods points to the importance of alcohol as a major cause of crashes.

These relative risk models (curves) are derived from point estimates (i.e. the risk of a crash at that exact BAC rather than a cut-off) of risk at numerous BAC levels and should not be confused with cut-off levels. I was able to estimate the unadjusted odds ratios (ORs)[1] of crashes for several BAC cut-offs for the Blomberg et al (2009) study from more comprehensive data (see Blomberg et al., 2005). OR calculations for different cut-offs are compared to point estimates (see **Table 10**). The ORs for thresholds are much higher than the relative risk model. For example, the unadjusted relative risk point estimate of being in a crash at .08% alcohol is 1.57 but the OR for this cut-off is 6.6. The reason for this difference is the cut-offs represent the **average** of all case-control comparisons above the cut-off and zero BAC, whereas the modelled data represent the risk at a particular BAC point.

Table 10: *Unadjusted Odds Ratios (OR) for crashes at different BAC cut-offs compared with modelled point estimates, derived from Blomberg et al (2005) data*

BAC	Cut-off Unadjusted OR	Unadjusted Point OR Estimates
Any positive BAC	1.8	
.02	2.4	.87
.04	3.4	.92
.06	4.7	1.13
.08	6.6	1.57
.10	5.7	2.37
.16	17.0	9.48
.24	24.1	21.92

From this data, it can be seen that from a public health perspective, a low per se BAC threshold, such as .02% alcohol, could still have a beneficial effect in terms of reducing alcohol-related crashes if it also deters drinking at the higher levels. There is scientific evidence that indicates very low BAC cut-offs can be effective (Mann et al., 2001). The other issue is whether per se levels at such levels are fair. When examining the relative risk model point estimates, it can be seen that drivers at BAC levels below .05% are not a crash risk and on the whole do not represent a safety risk[2]. Per se laws with cut-offs below .05% are more defensible from a public health perspective than a criminal justice perspective but are really aimed at those above a threshold of .05%, not those between .02% -.05%. As previously

---

[1] Adjusted ORs werecould not possible to calculatebe calculated based on the data available. The purpose of this example is to demonstrate the difference between cut-offsBAC cut-offs and those based on modeled risk curvespoint estimates. The data includes recovered hit-and-run crashes.

[2] Table 10 only shows unadjusted OR estimates; however, the same conclusion can be drawn from adjusted point estimates with 33 covariates (Blomberg et al., 2009).

mentioned, many safeguards protecting individual rights are enshrined in criminal code offences. However, similar rights are not enshrined in simple traffic-safety laws. Driving a car is not a right; it is a privilege. Therefore, laws can be implemented within traffic-safety codes that may not be fair to some individuals, but are best for protecting public health at large. Fines for speeding are an example of this principle. Not all drivers who are caught driving over the speed limit are necessarily a traffic-safety risk, but lower speed limits overall do save lives. A greater issue arises where drivers within a low range of BAC levels such as between .02% and .05% are prohibited, when the preponderance of scientific studies have not shown such drivers are a safety risk. As mentioned previously, the final decision on where to set the threshold for impairment depends in part on a trade-offs between fairness based on the proportion of false positives, the simplicity of the message and the public safety benefits. If an intervention targets mostly those who do not represent at safety risk, then it is difficult to justify from any perspective.

The risk of crashes below .05% alcohol

Some authors have suggested that per se limits below .05% are justified based partly on observational studies that found elevated crash risk at these levels (Mann, 2002). The studies presented here do not show an elevated risk of crashes at these low levels.

An interesting finding from most case-control studies of crash risk and BACs is that point estimates below .05% alcohol are below 1, suggesting a protective influence of alcohol. This protective effect was attributed by Allsop to a phenomenon called "Simpson's paradox" (Peck et al., 2007), a situation where the overall trend is the opposite in different subgroups. When controlling for one or more variables, the trend reverses itself. Based on Allsop's theory, the protective dip is not real. Personally, I am not persuaded that the dip is a reflection of Simpson's paradox. One reason is that Blomberg et al., (2009) controlled for 33 variables in a second model, plus matching in the design of the study and the dip still remained at up to .05% alcohol. The dip only disappeared in this study when the authors ***imputed*** BAC levels for missing hit-and-run drivers from those who were caught. Imputation is a procedure where BAC concentration levels are estimated for unknown drivers in crashes (i.e. missing data). Although many hit-and-run drivers who are later apprehended likely left the scene to evade detection of their drinking, it is also reasonable to expect that those who successfully evaded detection (i.e. never apprehended) were more likely to be sober than those who were caught. Drinking does make performance worse. The imputation procedure substantially elevated the risk of a crash at every BAC level. For example, the imputed RR for a BAC level of .25% increased from 20.29 (no covariates) and 26.60 (33 covariates) to

153.68 (33 covariates plus imputation for hit-and-run drivers), which is a very massive increase, especially considering the increase is due to an imputation procedure of unknown BACs among hit-and-run drivers. Overall, while this study provides concrete empirical evidence that hit-and-run drivers who are eventually caught are more likely to have been drinking than drivers who remain at the scene, the true prevalence of drinking in all hit-and-run drivers is still unknown. The prevalence of drinking is likely lower for hit-and-run drivers not caught than those caught, based on the reasons given above.

The more likely explanation for the dip in risk below 1 is that those who drive after one or two drinks tend to be social drinkers who take extra precautions while driving, possibly lowering their risk. Regardless of whether the dip is real or not, its magnitude, or effect size, is small. However, the dip certainly provides no justification that drivers at these low BAC levels represent a meaningful crash risk.

Some research has concluded that certain subgroups of drivers with BACs under .05% represent a significant crash risk. One study frequently referenced (Zador et al., 2000) conducted a secondary analysis of the Fatal Analysis Reporting System (FARS) data where BACs were taken from some fatally injured drivers, compared to those not in crashes derived from a 1996 National Roadside survey of drivers. They found that involvement in fatal single-vehicle and all fatal crashes was significantly related to BACs between .01–.019 for male drivers between 16 and 20 years old and to all drivers with BACs between .20–.49. The calculated RR for a fatal single-vehicle crash is an astonishing 15,559.85 for male young drivers with BACs over .15%, over 100 times greater than estimated by Blomberg et al. for drivers at a higher cut-off of .25 % alcohol. The magnitude of these effects and the absolute differences with the data from Blomberg et al. (2009) suggest that there are substantial biases in the Zador study. A closer inspection of the methods reveals that BACs also were imputed for many drivers (estimated by the National Center for Statistics and Analysis (2017) at about 38%) because actual BACs were not taken. Imputations like these create bias toward a stronger relationship than reality because BAC tests are more likely to be ordered for those suspected of drinking[1]. Data for this study likely has considerable selection bias, described earlier, where differential methods are used to collect data from cases (fatally injured drivers) and controls leading to erroneously high risk ratios. However,

---

[1] See section in this chapter, EPIDEMIOLOGICAL GUIDELINES FOR VALID FINDINGS, (1) Biases for more details on this limitation.

the magnitude of the risk ratios found in the Zador et al. studies are extreme and the methods are biased. This type of study should not be used to develop policies.

Datasets like the FARS and the European Driving Under the Influence of Drugs, Alcohol and Medicines (DRUID, European Monitoring Centre for Drugs and Drug Addiction, 2012) databases[1] are useful for understanding the issues but are plagued with inconsistent data collection approaches across jurisdictions and other biases that make findings misleading. Selection bias is a major drawback with the databases on fatalities is as different methods are often used to collect data from control subjects, previously discussed in this chapter. Non-***response bias*** among control subjects but not cases is a major limitation. This type of bias produces elevated ORs.

CANNABIS

In 1999, Bates and Blakely published a review article on cannabis and motor vehicle crashes. These authors concluded, "There is no evidence that consumption of cannabis alone increases the risk of culpability for traffic crash fatalities or injuries for which hospitalization occurs, and may reduce those risks" (Bates and Blakely, 1999; p. 231). At that time, there were three fatal injury studies on cannabis and driving, and all of them indicated a reduced risk of culpability for traffic fatalities. One major shortcoming of these earlier studies was a failure to separate carboxy-THC from THC. Since 1999, there have been several review papers on cannabis and crashes. A major limitation of many of these reviews is that the definitions of cannabis use were diverse. The review papers that did not separate the studies of blood, oral fluid tests and urine tests into groups for analyses, are reviewed directly below. Review papers that did separate groups are reviewed under the different biological specimens later on in this chapter.

Li et al. (2012)—This study was a meta-analysis on cannabis with crash risk. They combined findings with very diverse measures of cannabis use: one study with blood tests, two studies with urine tests and five self-report studies of cannabis use (not necessarily in relation to crashes). This latter group included two studies of marijuana use in general (i.e. with an unknown connection between a crash and cannabis) and three studies of marijuana use within three hours before a crash. The combination of these studies with diverse

---

[1] The DRUID data base includes several data sources where individuals were given drug tests (blood or oral fluid) including driver fatalities, roadside surveys, and hospital surveys.

assessment measures into a meta-analysis is not appropriate and makes objective generalizations impossible. Rogeberg and Elvik (2016) also found the selection criteria from the Li et al. study hard to rationalize. Furthermore, the Li et al. (2012) study did not control for alcohol consumption or any other potential confounders, such as sex and age, in their analyses. Since these flaws are major limitations, the findings from this study are given no weight.

Elvik (2013)—This review article is very extensive, with 66 studies of the relationship between drugs (not just cannabis) and crash risk. This is a useful paper due to the breadth of studies found, the author's statistical approach and concern with issues of confounding and causality. The author identifies laboratory analysis of blood samples as the most reliable indicator of recent drug use, and notes that metabolites of cannabis can remain in urine for a long time. As a result, in many studies that assessed cannabis use from urine tests, it is not clear that drugs were actually used shortly before driving. Furthermore, the author notes that estimates of risk tend to be higher in methodologically weaker studies than stronger ones. In fact, in the abstract, the author notes that most studies did not control well for potentially confounding factors, thus significant relationships should not be interpreted as causal. These authors estimated an OR of 1.16 for cannabis-related crashes from biological evidence (including blood, oral fluids and urine) but did not provide separate estimates for THC in blood. The low OR from this review may be related to inclusion of studies that used both more accurate blood tests and less accurate urine tests for detecting impairment. As mentioned previously, blood tests for THC concentrations are accepted as the most valid measure of impairment. Therefore, it is worthwhile to examine observational studies with blood samples that detect THC.

Gjerde et al. (2015)—This review paper was extensive, with 36 studies on cannabis. The authors also reviewed crash risk for several other types of drugs (benzodiazepines, opioids, stimulants and antidepressants). The authors do identify the type of samples used in each study but do not separate out their estimates of risk with blood tests versus other biological tests. They estimate the range in ORs for cannabis exposure and crashes between 1 and 4.

White (2017)—In this report, numerous observational studies and methodological issues related to cannabis and crash studies are examined. He concludes that the ***odds ratio*** for cannabis and crashing is unlikely higher than 1.30 and that there is no compelling evidence of a dose-response relationship.

Hostiuc et al. (2018)—The authors of this study identified 24 studies that examined the relationship between cannabis and unfavourable traffic events. They reported an odds ratio of 1.97 for studies with blood tests, a weak effect size. However, several of the studies that used blood tests incorporated mixed measures that typically included oral fluid tests in the control subjects (for example, Gjerde et al., 2013; Li et al., 2013 and Movig et al., 2014). Mixed measures such as these create bias. The findings are uninterpretable due to the poor relationship between THC in blood and oral fluids, described in more detail in Chapter 3.

Crash risk studies of THC in blood

As mentioned in Chapter 3, blood tests for THC are currently accepted as the best biological sample for detecting likely impairment. There have been observational studies of crash risk for oral fluid and urine tests, briefly reviewed at the end of this chapter. Several case-control or culpability studies have been conducted on THC in blood and crash risk. Below I highlight one of these studies (Drummer et al., 2004) and a meta-analysis review paper that includes all the higher-quality studies with blood tests (Asbridge et al., 2012). Both of these studies are strong contributions to the field.

Review papers of THC in blood and crashes—In two comprehensive studies, both laboratory studies and observational studies for cannabis risk in driving were reviewed and the authors suggested THC blood concentrations that might constitute impairment (Grotenhermen et al., 2007; Hartman et al., 2013). The paper by Grotenhermen et al. (2007) is a summary of deliberations from a panel of international experts, who wrote a more comprehensive report two years earlier (Grotenhermen et al., 2005). The research from this team was divided into laboratory studies and epidemiologic observational studies, similar to how I have presented material in this book. This group put much more emphasis on laboratory studies, described in Chapter 5, acute symptoms and duration of impairment/cannabis. I concluded in this section that the authors generalized findings from studies without blood tests to draw conclusions on a possible per se limit between 3.5 and 5 ng/mL THC in whole blood. This kind of generalization violates basic principles of science, not just the discipline of epidemiology. For the observational studies, they relied heavily on the study by Drummer et al. (2004) described above and in more detail below in the section "Comparisons of risk for alcohol and cannabis."

Hartman et al. (2013) reviewed epidemiological field studies that used biological tests, primarily blood tests for THC, to examine the association with crashes. Many of the studies were included in the review by Asbridge et al. (2012), cited above. The conclusion that THC concentrations of 2–5 ng/mL are associated with substantial driving impairment is not justifiable by the methods. First, none of the studies reviewed actually reported THC levels in this range, other than Grotenhermen et al. (2005) based on a secondary analysis of data by Drummer et al. (2004). This analysis actually shows a protective effect in this range (see Figure 7, below). The error in reasoning appears to be related to understanding the meaning of OR cut-offs, described in more detail above in relation to alcohol (see Table 10, above). Essentially, if low cut-offs are used, comparisons are based on the average of everyone above and below the cut-off. If there is a dose-response relationship, the OR can be misleading, including those that exhibit a protective effect as well as a risk factor. ORs based on cut-offs should not be confused with point estimates. For example, in the data from Blomberg et al. (2005), an OR for any positive alcohol test is 1.8, but point estimates are below 1 at BAC levels of .04% and below. In the review by Hartman et al. (2013) all of the studies except Drummer et al. (2004) reported very low cut-offs of 1 ng/mL or below in blood so we cannot draw conclusions about crash risk for levels between 2–5 ng/mL from these other studies.

Drummer et al. (2004)—In my opinion, this study is an important contribution to our understanding of the relationship between crash risk and cannabis use, although the study has limitations. Drummer published some early studies where he did not find a significant relationship between culpable risk and cannabis, as cannabis positives had lower risk for culpability. As mentioned earlier, a shortcoming of these early studies is that blood tests often did not distinguish between the metabolite of carboxy-THC and THC (Bates and Blakey, 1999). In this 2004 study, Drummer specifically aimed to distinguish between carboxy-THC and THC. He found that THC was associated with increased culpability among fatally injured drivers but carboxy-THC was not. The culpability risk curve for THC levels was modelled and presented in the paper by Grotenhermen et al. (2007). This modelled risk curve, presented below (see **Figure 7**), shows a protective effect at concentration levels below 5 ng/mL similar to the protective dip at low BAC levels below .05% found in case-control studies for alcohol, with the risk curve for alcohol also shown, based on unadjusted crash risk for alcohol (Blomberg et al., 2009). More details are provided in the next section of this book.

Figure 7: *Comparison of crash (responsibility) risk curves for BAC and THC levels*

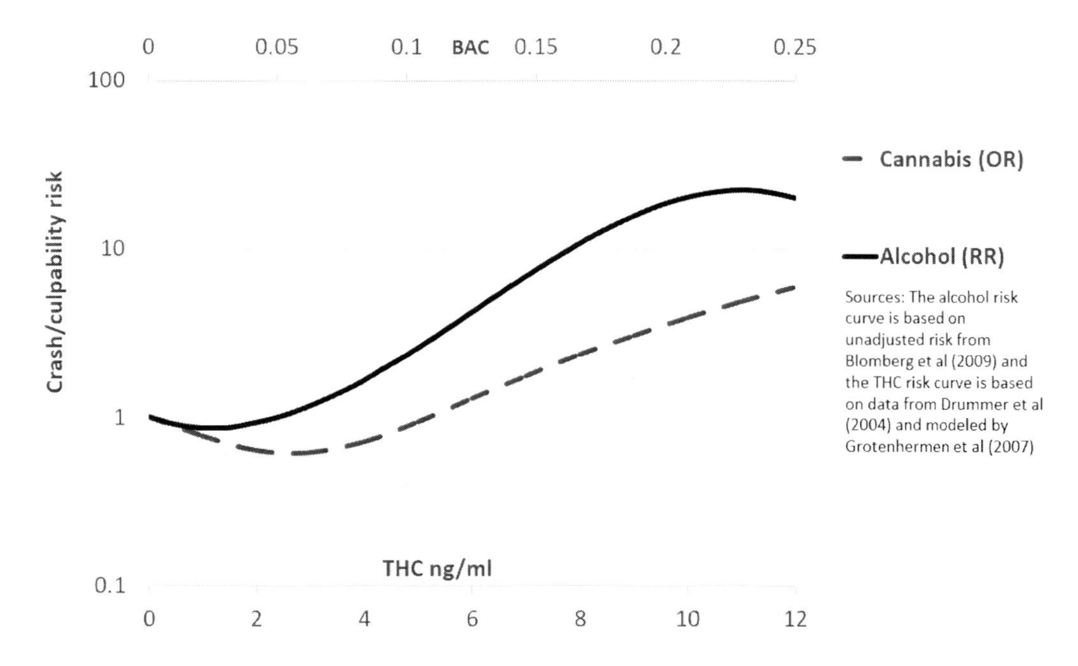

An increase above one OR begins at about 6 ng/mL THC. Drummer et al. (2004) reported an odds ratio of 6.6 at a cut-off of 5 ng/mL. It should be noted the risk curve from the Drummer et al. (2004) data are modeled smoothed estimates an not based on actual point estimates as done by Blomberg et al. (2009). Only 58 drivers had positive THC samples in the Drummer study, too few to recommend a per se limit, and additional studies have flaws or that preclude a per se recommendation (Grotenhermen et al., 2007). According to this model, 6 ng/mL THC could be considered equivalent to .08% alcohol in terms of crash risk. The effect sizes between THC in blood and being culpable were much greater than other studies, as illustrated in findings of review articles described below. The effect sizes in the Drummer study may be over-estimated, as explained by White (2017).

Asbridge et al. (2012)—In this meta-analysis, the authors limited their inclusion criteria to empirical studies of the risk of a motor vehicle collision while under the influence of cannabis (defined by self-reports of driving within three hours after use or positive THC levels from blood tests) and excluded all studies with urine and oral fluid tests (Asbridge et al., 2012, p. 3). An important feature of this review is that only medium and high-quality studies (based on methodological characteristics to reduce bias) were included. They assessed four studies as high quality, all culpability studies (Drummer et al., 2004; Laumon et al. 2005; Terhune et al., 1992; Terhune & Fell, 1982). Taking all the studies combined, acute cannabis consumption was associated with an increased risk of collisions compared to unimpaired drivers (OR = 1.92). Asbridge et al. (2012) noted a major bias was concurrent use of alcohol and cannabis, which, if left uncontrolled, would overestimate the effect of cannabis. To address this bias, Asbridge et al. eliminated all subjects with positive BACs from their analysis (2012). A limitation of their review is that basic confounding variables of age and sex were not controlled for in their analyses, which means the ORs reported could be on the high side. For studies with tests of concentration of the active THC in blood samples, the odds ratios for a collision/culpability ranged from .82 to 4.4 (Asbridge et al., 2012), with six of nine studies finding significant relationships. I see this meta-analysis as the most important contribution to our scientific understanding of the relationship between the acute effects of cannabis and crash risk/responsibility. Rogeberg and Elvik (2016) re-analysed the studies from Asbridge et al. and conducted a meta-analysis with 21 studies. Their analysis led to lower OR risks of below 1.5 for cannabis.

Summary—Although an overview of the research evidence shows that being under the influence of cannabis is related to crash risk, this relationship is not as strong as the relationship between alcohol and crash risk. The preponderance of the evidence indicates that the odds of being in a crash or responsible for testing positive for cannabis is less than 2, a small effect size. There are several limitations of the research. None of the studies on cannabis have established a compelling dose-response relationship between THC concentrations and crash/culpability risk, comparable to the studies on alcohol. Rather, researchers in most studies chose a low cut-off for THC, and dose-response relationships were not examined properly. By comparison, if researchers used a very low threshold for drinking, such as BAC positive versus negative, we would arrive at ORs of similar magnitude. I calculated a crude OR of 1.8 for crash risk for positive BACs using the detailed data from Blomberg et al. (2005). The overall OR is low because drivers with BACs under .04% alcohol are less likely to be in crashes and those just over .04% BAC have much lower risk than those with very high BACs. Yet, because a much greater proportion of the population drive under the influence of alcohol than cannabis, most studies of alcohol have found statistical significance at even low cut-offs for alcohol.

None of the above-mentioned reviews proposed a cut-off value of THC in blood that might correspond to impairment. A major limitation of existing research is insufficient data to present a dose-response relationship between THC concentration levels and crash risk. Estimates of the OR for the acute effects of cannabis and crashes are under 2, based on low thresholds of 1 or less, a weak effect size (see Table 9). The low ORs found in review papers may be partially due to comparing THC positives versus negatives rather than choosing a higher threshold that might constitute impairment.

COMPARISONS OF RISK FOR ALCOHOL (BREATHALYZER) AND CANNABIS (BLOOD TESTS)

The epidemiological evidence shows that collision risk increases noticeably at .05% alcohol (adjusted OR=1) and increases exponentially with higher BACs, with RR estimates of at least 25 for BACs above .20%. By contrast, for risk related to active THC in the blood, the results from higher-quality studies yield much lower risk estimates and dose-response relationships. Overall, the risks from cannabis appear much weaker than for alcohol. As well, the relationship between the acute effects of cannabis and crash/culpability risk is much less conclusive than for alcohol.

Figure 7 above illustrates a comparison of the modelled crash risk for BACs and THC levels in blood. In creating this figure, I have used exact numbers from the best crash unadjusted crash risk for alcohol (Blomberg et al., 2009) and for cannabis, an estimated OR curve modelled by Grotenhermen et al. (2007), based on a high-quality study by Drummer et al. (2004). The studies are not perfectly comparable for several reasons. Drummer et al. (2004) conducted a culpability study of fatal drivers whereas Blomberg did a case-control study of drivers in crashes of all types. This difference could create an upward risk bias for cannabis due to differences in research design (i.e. culpability versus case-control design) (see Rogeberg and Elvik, 2016) and because cannabis use may be more strongly associated with fatal crashes than non-fatal ones (similar to findings for alcohol). Another source of bias could be temporal proximity between the event and test. Metabolism of THC in blood halts at time of death and additional THC is released into the bloodstream post-mortem as fat cells breakdown. Therefore, the overall THC concentration levels recorded in deceased individuals by Drummer et al. (2004) are likely higher than one would expect from a case-control design with blood samples, as with live drivers, THC concentration levels would be greatly reduced in the interval between crash and blood test. The Drummer et al. (2004) study found a high OR compared to other similar studies (see Asbridge et al., 2012). These factors suggest that the ORs found in the Drummer et al. (2014) study may be biased upward. One final and important limitation of the Drummer et al study (2014) is that this study only had 58 cases positive for THC only, far too few to draw conclusions with confidence. Overall, I have taken a conservative approach in comparing the two modelled risk curves. The alcohol risk curve may be higher in reality and the THC risk curve may be lower in reality for the reasons described in this paragraph.

The comparison does indicate that the absolute risks of crashes are greater for alcohol than cannabis taking into consideration the typical amounts of alcohol or cannabis that people consume before driving. Sewell et al. (2009) arrived at a similar conclusion. Comparison of the models shows the risk curve becomes greater than 1 for alcohol at about .05% and at about 6 ng/mL for cannabis. Below 6 ng/mL is a weak protective effect. This data, given all the caveats and potential biases described above, provides the only concrete real-world evidence that a per se limit of 6 ng/mL may be comparable to .05% alcohol.

CRASH RISK STUDIES WITH ORAL FLUID AND URINE TESTS FOR CANNABIS

Oral fluid (OF) tests

With respect to oral fluid tests, I am only aware of one study using OF tests for both cases and controls (Compton & Berning, 2015; a secondary analysis of a data of a study conducted by Lacey et al., 2016). This study aimed at replicating the strong design features by Blomberg et al. (2009), except OF and blood tests were taken for detection of substances rather than Breathalyzers for alcohol. Neither cannabis nor any other drug was significantly related to crash risk when controlling for simple demographic variables and alcohol consumption (Compton & Berning, 2015). It should be noted that the cut-off for cannabis positive was extremely low at 2 ng/mL in oral fluids which may explain why significance was not found. I suspect significance could be achieved with analyses of THC concentration levels as a continuous variable, which was not done. Only blood alcohol content above .05% alcohol was significantly related to crashes in statistical tests, a finding that has been replicated in numerous other studies. Although some studies have used oral fluids for one group, usually the controls, this approach is not sufficiently accurate as conversion factors for individuals from one type of sample to another are unreliable.

Urine tests

Table 11 shows some key characteristics of road-crash studies that used drug tests of urine for the inactive metabolite of THCCOOH or oral fluids for THC. The right-hand column of the table shows the statistical associations conducted by the authors using statistical tests that controlled for confounders, if conducted. ***Non-significant*** (ns) means that no statistical relationship was found between positive urine tests and crashes and therefore the magnitude of the ORs should be ignored. Only two papers found a statistical relationship between positive tests and crash risk (Dussault et al., 2002; Brault et al., 2004); however, these two studies are both based on the same dataset and therefore should be treated as a single study (Brault et al. is the larger dataset). Brault et al. (2004) did not find a significant relationship between testing positive for cannabis and responsibility for the crash, and these data based on a culpability design were not subject to selection bias (although the case-control portion was biased, as described below). Among the five other studies using urine tests, none were statistically significant.

Table 11: *Characteristics of crash risk studies using urine tests to detect carboxy-THC*

Author & date	Study type	OR (p level)
Brault et al. 2004	Culpability	1.2, ns
Brault et al. 2004[1]	Case-control	1.6, p <.05
Lowenstein & Koziol-McLain, 2001[2]	Culpability	1.1, ns
Marquet et al. 1998	Case-control	ns
Soderstrom et al. 2005	Culpability	1.18, ns
Woratanarat et al. 2009	Case-control	0.78, ns

1. Odds ratios adjusted for age, gender, hour and day
2. Odds ratio adjusted for age, seatbelt use, sex, time of day/week. Alcohol was not included.

Carboxy-THC is also detectable with blood tests, and studies with blood tests also show carboxy-THC is not a good compound to test for impairment. Although carboxy-THC has a shorter detection period in the blood than urine, it remains detectable in blood well beyond its psychoactive effect period (Huestis et al., 1992). Drummer et al. (2004) completed a well-controlled responsibility study in Australia and found no significant difference between cases and controls for the presence of carboxy-THC without THC in blood. Ramaekers et al. (2004, p. 111, Table 1) present six crash-risk culpability studies that tested for carboxy-THC in blood. No significant ORs for carboxy-THC were found in five of these studies; only one study showed a significant protective effect (Terhune et al., 1992).

These observational studies consistently show that detection of carboxy-THC in urine is not related to performance deficits, as determined by crash or responsibility risk. The single study conducted regarding oral fluids and crash risk shows no statistical relationship between testing positive for any drug and crashes. There is ample substantiation of my position. The studies are consistent in their failure to find statistical significance. Similar findings show no reported relationship with tests of carboxy-THC in blood (Ramaekers et al., 2009). The review of the above studies supports my original conclusions. There is substantial evidence that carboxy-THC is not related to crash risk.

Brault et al. (2004)—This study is based on Quebec data included both a case-control and culpability research designs, also reported by Dussault et al. (2002). The case-control portion of the study has two sources of selection bias. Selection bias is a distortion of an association that can be found: (1) in the cases (acknowledged by the authors), and (2) in controls, due to non-participants being more likely to have used drugs than participants since consent to testing was required for controls but not cases. Given the choice, individuals who had recently used drugs would be less likely to agree to a drug test than those who had not recently used drugs. These two types of bias will have the effect of increasing the magnitude of statistical relationships for two reasons. First, in terms of selection bias among the cases, the authors note that urine tests were much more likely to be taken for young drivers and men in fatalities (see Brault et al., 2004, Table 2). Since both men and 16–24-year-olds are more likely to use illicit drugs (including cannabis) than others (Statistics Canada, 2012), this bias would also lead to a greater proportion of drug users (i.e. an unrepresentative sample) in the driver fatality group, and have the ultimate effect of artificially inflating the ORs. Second, the existence of selection bias in this case-control component of the study (in fact, only 49.6% of controls provided a urine sample) is also illustrated by the authors' responsibility analysis that was free of these biases (Brault et al., 2004). Statistical significance was **not** found in the responsibility analysis. An additional confounder is that those who use drugs may simply be riskier drivers. We do not know whether risk taking or drug use is a better explanation of collision risk. The authors acknowledge these methodological limitations.

Studies with mixed biological samples

Several studies of collision risk and drug detection used different biological specimens for those in collisions versus controls (Gjerde et al., 2011; Movig et al, 2004; Hels et al., 2013; Bogstrand et al., 2012; Li et al., 2013). Since the correlations between drug concentrations with different biological samples are unacceptable, this mixed approach introduces considerable error and bias into the findings. Therefore, these studies are not reviewed here.

Evaluation studies

Numerous studies have assessed the impact of per se laws for alcohol (Ross, 1984; see Mann et al., 2001 for a review). The preponderance of evidence indicates that for alcohol-related crashes, reductions are sometimes only short-lived after new per laws are introduced. By contrast, a recent review article on cannabis legislation concluded that no study has been published that evaluated the impact of laws against driving under the influence of cannabis (Watson & Mann, 2016; p.152).

This approach is based on the principle of deterrence. The cornerstone of deterrence is punishment. According to deterrence theory, if undesirable acts are punished, then people will be less likely to engage in those behaviours in the future. This type of deterrence is referred to as specific deterrence. Also, general deterrence is postulated a punishment that is expected to have a beneficial effect on the larger population, who will be less likely to engage in behaviours based on the likelihood they could be punished. There are three components of deterrence related to effectiveness. If the punishment is certain, swift and severe, people will be less likely to engage in those behaviours. For some types of offences there is convincing empirical evidence that the certainty of punishment has a greater deterrent effect than increasing the severity of punishment (Wright, 2010; Nagin & Pogaesky, 2001).

Figure 8: *Moving average monthly alcohol-related fatal collision rate per 1,000,000 licensed drivers in BC, 1996–2012*

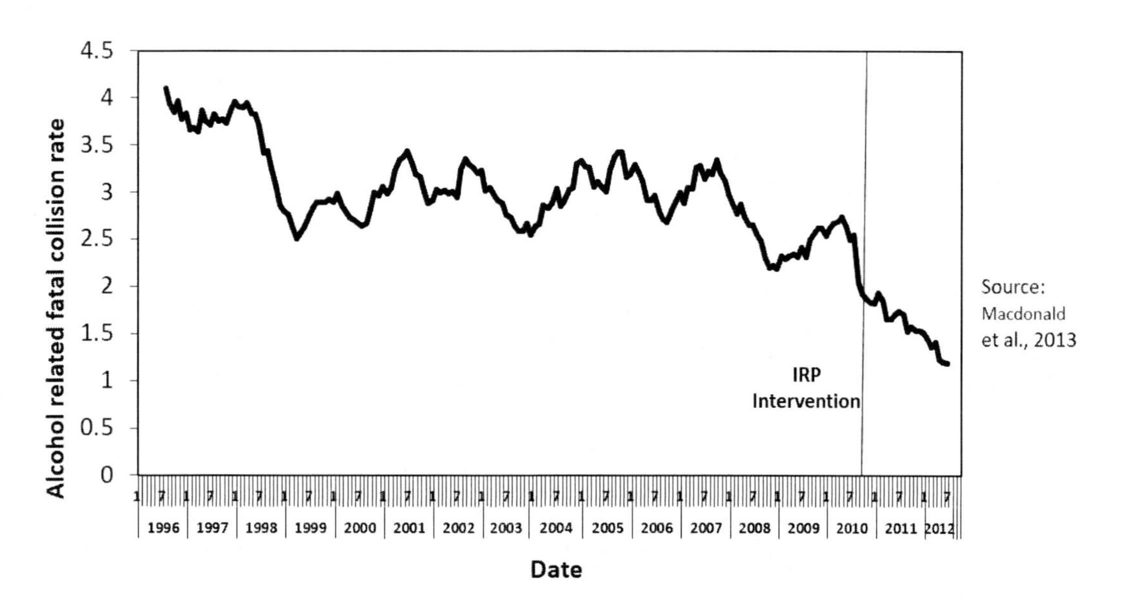

In one study in BC, Canada, we evaluated a provincial law, "Immediate Roadside Prohibition (IRP)," where drivers with BACs over 0.5% received immediate impoundment of their cars (Macdonald et al., 2013). This law, aimed at providing swift intervention, showed promising positive effects at reducing three types of alcohol-related crashes. According to our analysis of traffic fatalities, the rates of fatal crashes significantly declined after introduction of the new law in 2010 (see **Figure 8**). Findings similar to these for new legislative initiatives, such as new or lower per se limits, have been reported in several other studies (Ross, 1984; Mann et al., 2001).

In the U.S., police must have reasonable grounds that a driver has been drinking before demanding an alcohol breath sample (a process called ***selective breath testing***). Legislation in Canada (December 18th, 2018) allows ***random breath testing*** of drivers. Random testing allows police to stop drivers at random and ask them all to provide a breath sample without any reasonable suspicion of drinking, a process also legal in Australia. Some have raised concerns that the law could be used to unfairly target visible minorities as breath tests can be ordered without suspicion (Dickson, 2018). Supporters of this law note that a certain proportion of drivers are able to deceive police into thinking they are not over the limit and evade detection (Solomon et al., 2011). Thus, the certainty of being apprehended is increased if you are stopped by the police. As well, some people feel it is actually fairer since every driver is treated in the same fashion and no questions about prior drinking need to be asked. However, the likelihood of being caught for impaired driving will still remain low after the introduction of such legislation, making it unlikely that it will have a measurable long-term impact on alcohol-related arrests or crashes. Overall, empirical research has not shown that truly random breath testing will produce safer roads than selective breath testing (Elder et al., 2002; Shultz et al., 2001).

CONCLUSIONS/DISCUSSION

For the case-control/culpability studies, research shows much stronger relationships between alcohol and crashes than cannabis and crashes. There is some evidence that blood tests can be used to determine impairment for cannabis; however, thresholds (i.e. cut-offs) to classify individuals as impaired have not been established for blood or any other biological sample.

For alcohol, the epidemiological evidence shows strong dose-response relationship between BAC level and crash risk, increasing noticeably at .04 -.05% alcohol. By contrast, for active THC in the blood, too few studies exist to provide a dose-response relationship similar to that of alcohol. Collision risk and the results from higher-quality studies are too inconsistent to draw strong conclusions; however, with great confidence, we can say that alcohol is much more dangerous than cannabis, both in absolute terms and from a burden of disease perspective.

It is unlikely that existing and proposed laws worldwide based on deterrence will be effective in reducing cannabis-related crashes. There are substantial differences in the evidence base for per se laws with alcohol and cannabis. I believe these differences have a fundamental impact on the effectiveness of deterrence-style laws. First, if the confirmatory cut-off levels remain low, as they are today in most countries, the likelihood of detecting those *who are impaired* (i.e. a meaningful safety risk) is low and therefore most who are not a risk will be punished. The inaccuracy of the biological tests for cannabis in detecting a threshold that corresponds to a meaningful safety risk or impairment will undermine enforcement practices, especially since cannabis use has become legal. In North America, laws within the criminal code should be perceived as fair and just. Low thresholds mean that the laws could be challenged in courts for fairness and over time, police may become less vigilant in enforcing such laws. Although a high cut-off with THC blood tests likely could be used to reasonably conclude impairment, a high cut-off is only feasible if blood samples are taken immediately when impairment is suspected, and only when cannabis was smoked, not ingested. An immediate blood sample appears to be unacceptable under current Canadian laws. It is even a challenge to obtain a blood sample of unconscious hospitalized drivers for alcohol analysis (Chamberlain & Solomon, 2010). Second, unlike the Breathalyzer for alcohol, no drug tests for portable tests for cannabis exist which have the same degree of accuracy as breath tests. Most importantly, however, research evidence has not convincingly shown that THC blood tests at cut-offs of 5 ng/mL or below indicate a reasonable likelihood of impairment.

BAC thresholds below .05% alcohol, as exist in Scandinavians countries, can also have a beneficial effect on alcohol-related crashes. These low limits send a message not to drive within at least an hour after having a single drink or not to drive for several hours after consuming several drinks. Even with these low limits, those caught minimally will have consumed alcohol within a short time frame before driving or consumed a very large amount within the past day. The message is simple to understand. Although those over the limit on average of will have increased risk of crashes compared to those below a low limit, many of those caught above a low may still not represent a real safety risk. For cannabis, the time frame in which driving must not occur is much longer at levels below 5 ng/mL for a THC blood test than for alcohol at .02%, perhaps up to several days for daily users, and a low limit is not simple for users to understand.

CHAPTER 8

Conclusions

MYTHS AND TRUTHS

Data presented in this book shows there are many misconceptions about cannabis use and impairment. I have not attempted to provide a comprehensive examination of all the issues. Rather I have focused on more major works in the field that I believe have had an impact on legal policies, but their conclusions are not justified by the methods used. A summary of these myths and truths is provided in **Table 12**. In the sections that follow, I highlight some of these scientific studies from which these myths originated.

Table 12: *Some common myths and truths about cannabis, and how to measure impairment*

Myths	Truths
The validity of THC cut-offs in blood as an indicator of impairment has been established.	No validity studies of the relationship between THC in blood and performance, using an epidemiological approach, have been published. Therefore, no firm conclusions can be made on the accuracy of any blood THC cut-off for impairment. Higher levels of THC in blood are related to a greater likelihood of impairment.
A simple conversion factor from THC in serum or plasma can be used to estimate an individual's whole-blood THC level.	Substantial variability exists between individuals of serum/plasma-to-whole-blood ratios. Since this variability is large, an average cannot be used to estimate an individual's whole-blood THC level.

Myths	Truths
Subjects given the same amount of THC will have similar THC concentrations at the different time points.	There are substantial differences between individuals based on blood and oral fluid samples.
Concentration levels of THC in blood are the same as THC levels in oral fluids.	Concentration levels of THC in oral fluids greatly exceed concentration levels in blood. Oral fluid/whole blood ratio is likely between 20 and 60, with higher ratios noted shortly after use.
Oral fluid tests or urine tests are equally good as blood tests for detecting impairment.	Blood is the best biological sample to detect impairment from either cannabis or alcohol.
THC concentrations in blood will be greater when cannabis is eaten than when smoked.	THC concentrations are much greater in blood within the first hour after smoking compared with the same amount eaten.
THC in blood is eliminated at a constant rate, like alcohol.	THC is eliminated from the blood rapidly in the first hour after smoking and slowly thereafter. THC is eliminated more consistently when ingested.
Impairment and performance deficits are the same thing.	Impairment is a legal term that states performance deficits meet a particular threshold for legal penalties. Performance deficits refer to any decrement.
Immunoassay screening tests detect the same compounds as confirmatory tests using chromatography and spectrometry analysis.	Immunoassays detect several types of compounds with similar molecular structures, but the combination and concentration levels of each compound is unknown. Confirmatory methods are used to detect a specific compound, such as THC.

Myths	Truths
THC is the only compound or metabolite that is psychoactive in cannabis.	Cannabis contains over 400 compounds, several with psychoactive properties. Hydroxy-THC is a metabolite of THC that is highly psychoactive, perhaps even more potent than THC.
Oral fluid tests for cannabis have been validated by the Canadian police.	Oral fluid tests for cannabis have been validated under laboratory conditions. Oral fluid tests were pilot tested the under Canadian conditions by police, and 65% of the tests were administered outside the manufacturer's suggested operating temperatures. The validity of the tests is unknown at these temperatures.
THC levels in the blood correlate closely to THC levels in the brain, similar to how alcohol content in the blood correlates with alcohol content in the brain.	Although there is a positive correlation between cannabis in the blood and brain, the relationship is weak, whereas the relationship is very strong for alcohol.
Effects of cannabis are similar for different people.	Substantial variability exists among individuals for THC levels and performance deficits. Occasional users tend to have greater performance deficits than daily users following acute effects.
The acute effects of cannabis are equally harmful as alcohol.	Research shows that alcohol is more strongly related to crash risk than cannabis at typical amounts that people use when driving.
We know how to measure impairment in laboratory situations that is at a magnitude known to be associated with crash risk.	Impairment (meaningful performance deficits) for cannabis is not well defined.
Cannabis impairment lasts 24 hours.	The preponderance of research shows cannabis impairment up to four hours after smoking and up to 10 hours from ingestion.

A misleading aspect of many studies is an emphasis on measures of central tendency at the expense of measure of dispersion. An important example of this limitation is that studies have recommended per se limits from biological samples, typically THC concentrations in blood and performance. Not a single study was found that used an epidemiological approach to validate any cut-off level (for a possible per se limit), an approach that incorporates individual variability of THC cut-offs in relation to impairment. This research limitation appears to permeate the field of research of cannabis and crashes in other ways. Substantial variations exist between individuals in the absorption and elimination of THC at different time points, yet often summary statistics of central tendency are used to describe the relationship, which can give impression of a high degree of similarity between people. Another example is the assertion that an average serum/plasma blood THC concentration can be used to estimate an individual's whole-blood THC level based on an average ratio between the two. This assumption does not take into consideration the fact there is substantial variability between individuals for this ratio, and the ratio can change for a given individual at different points in time. Therefore, any estimates based on a conversion factor will have a high degree of error for a given individual.

Greater distinction and understanding by researchers needs to be made between effect sizes and probability values. Probability values are based on the sample size and the dispersion of the variables and indicate the likelihood that a relationship is due to chance based on exactly the same methods of collecting the data. Effect sizes refer to the actual strength of statistical relationships. Statistical significance (i.e. $p<.05$) is best thought of as necessary but not sufficient to conclude a relationship is meaningful. If a statistical test is not significant, then there is no point in looking at the strength of a relationship. In this book, I have defined effect sizes that I consider weak, moderate or strong based on different levels of measurement. The degree to which the magnitude of an effect size is considered appropriate depends on the purpose of the relationship. Numerous other epidemiological research issues were raised in this book, illustrating how failure to adhere to standard research protocols can culminate in erroneous findings. Issues such as bias, confounding, measurement error, effect sizes and the difference between correlation and causation were highlighted.

Origins of the myth that the safety risk for THC levels in blood between 3.5 and 5 ng/mL is comparable to .05% alcohol

A common myth is that THC levels at low whole-blood levels below 5 ng/mL represent a safety risk equivalent to .05% alcohol. One origin of this myth likely originates in a highly influential paper by Grotenhermen et al. (2007). The paper is co-authored by 11 top researchers on cannabis and performance, and published in a high impact journal, *Addiction*. The paper is well-written, transparent and provides substantial scientific language that gives the paper much authenticity. The authors' conclusion that 3.5 to 5 ng/mL THC in whole blood is comparable to BAC levels of .05% alcohol is not supportable by scientific conventions. The conclusion is drawn primarily from laboratory studies where no blood tests were conducted, a conclusion that cannot be justified unless direct comparisons are conducted between the two, and no studies are cited where these comparisons were done. A very common principle of research, not restricted to epidemiology, is that one should not draw inferences about the findings (i.e. generalizations) to a much broader population than the sample used in the study. For this reason, and based on contradictory empirical evidence highlighted in this book, this conclusion is a myth.

Researchers Hartman et al. also relied on similar laboratory studies and observational studies to arrive at the conclusion that "...recent smoking and/or blood THC concentrations of 2–5 ng/mL are associated with substantial driving impairment, particularly in occasional smokers" (Hartman et al., 2013; p. 478). These authors drew improper generalizations about blood tests and impairment. No studies were cited that even examined THC blood levels in this range. The reason for this conclusion may have been based on findings from some studies where cannabis-positive drivers were at significantly greater risk of crashes/culpability; however, from these results it cannot be concluded that 2–5 ng/mL represents substantial driving impairment. When cut-offs are used, risk (OR or RR) is determined by the average of all drivers above the cut-off, which can represent substantial variations when in doses. Assuming a dose-response relationship, those with very high THC levels will contribute more to the overall association than those close to the cut-off. Again, the conclusion by the authors is not justified by the methods used.

In another laboratory study, the authors "...concluded that serum THC concentrations between 2 and 5 ng/mL establish the lower and upper range of a THC limit for impairment" (Ramaekers et al., 2006; p. 114). This range would be much lower in whole blood. Although paired blood and performance tests from subjects were taken, the analyses were incorrect, likely resulting in a stronger relationship between THC levels and performance, yet still a very weak effect size. This weak relationship was then used to justify a low threshold per se limit. I believe that this study also contributed to the myth that drivers with blood THC levels between 2–5 ng/mL represent a safety threat on the road.

Simply put, the studies that have asserted impairment at these low THC blood levels do not conform to basic standards of science. The papers above have likely been read by legislators who do not have scientific backgrounds for understanding the flaws in these studies. The consequence is that laws may have been implemented based on methods that do not conform to accepted scientific standards.

Origins of the myth of 24-hour deficits from cannabis

Health Canada (2013) reports that the ability to drive or perform activities that require alertness may be impaired for up to 24 hours following a single joint. A report from the National Highway Traffic Safety Administration in the U.S. with expert advice from six toxicologists states that some investigators have demonstrated residual effects in specific behaviors up to 24 hours, such as complex divided-attention tasks (Couper and Logan, 2004). The World Health Organization (undated) reports "human performance on complex machinery can be impaired for as long as 24 hours" after smoking a moderate dose of cannabis.

The source of the claim of 24-hour deficits from each of these above-mentioned reports—just a few of many that exist in literature around cannabis impairment—is based on a single 1991 study by Leirer et al. The findings of this study have received an extraordinary degree of acceptance, but has multiple methodological issues, including a failure to randomize conditions, a failure to conduct a pre-test of the methods, and retention of unreliable measures. As illustrated in greater detail in Chapter 6, the preponderance of the evidence from studies with higher-quality methods indicates 24-hour performance deficits from the acute effects of smoking cannabis is a myth.

The reasons for myths

One might ask, "How could conclusions be made that are not justifiable from the methods?" There are likely several reasons for these myths; however, I do not think it is likely that these authors intentionally conducted research to mislead. The conclusions are likely driven from one overarching truth, i.e. cannabis impairs performance. Clearly, a public-health issue of cannabis-impaired driving has been identified and the desire to make recommendations to address this issue is great. This truth, along with the constant bombardment in the popularized literature of newspapers, magazines and television of the negative effects of cannabis, shapes opinions and inspires a desire for solutions. Researchers can be biased in their approach, not necessarily intentionally. This bias has long been recognized and disciplines have established principles of conduct, such as researcher blinding when carrying out experiments. Researchers are also under pressure to publish their research. As an academic all of my career, I have felt this need to publish, by both clearly stated management directives and more discreet social pressure from colleagues. Also, I have noticed that certain types of findings or conclusions, specifically those with significant results and conclusions with policy recommendations, are more likely to receive favourable reviews and be accepted for publication. This phenomenon, known as "publication bias" is accepted as real, and can contribute to a biased representation of studies in the literature.

In my experience, papers with large teams represent divergent expertise of different members of the team. The field of substance use research is highly multi-disciplinary with researchers trained in sociology, psychology, history, biochemistry, law, toxicology, pharmacology and epidemiology, to name some. Papers generally do not represent a consensus where every author has endorsed every sentence. Rather, multi-disciplinary teams are created and individual team members address different aspects of a topic, typically within their discipline. In addition, reviewers can sometimes miss important components of studies that should be revised because their discipline is incompatible with the discipline of the research. In any case, mistakes can be made and mandating that authors describe their methods allows researchers to independently assess whether conclusions drawn in studies are justifiable by the methods. In this book, I have examined studies from an epidemiological perspective. Researchers from other disciplines may not be trained in epidemiological methods and therefore flaws in the data collection and analysis approach may emerge through lack of knowledge of the discipline.

Alcohol and cannabis compared

Research has shown that the acute effects of cannabis cause performance deficits that increase safety risk while driving. The question of importance is: how do we best make the roads safer? For alcohol, the answer has been to first detect impaired drivers, then to administer sanctions against those who are a safety hazard. Per se laws in North America for alcohol are generally viewed as fair, as they target the subset of drinkers who are dangerous on the road. Furthermore, research shows per se laws are effective in reducing (but not eliminating) drunk driving. Many feel that we should adopt similar legislative penalties for those impaired by cannabis. Others caution that action that does not achieve its intended purpose and is harmful to some (i.e. false positives) cannot be justified.

Legislative programs should specifically target cannabis use that is a safety hazard rather than targeting cannabis users in general.

Major differences exist between cannabis and alcohol. Alcohol is much more widely used than cannabis. However, research indicates that at the quantities generally used by populations, driving under the influence of cannabis is less dangerous than alcohol, although both have safety risks. The pharmacokinetics of alcohol cannot be applied to cannabis. Alcohol is a simple, water-soluble drug, resulting in blood concentration levels which correspond very closely to performance levels. THC is fat-soluble, meaning it tends to bind closely to fat cells. THC concentration levels in the blood only correlate well with performance deficits within an hour after use when smoked, and THC levels can remain above 5 ng/mL in daily users over one day after use (Desosiers et al., 2014). Alcohol is eliminated from the body at a constant rate, whereas THC is variable at different points in time. Furthermore, the elimination process of THC from the body is much slower than for alcohol and provides a poor correspondence to impairment at low levels. Extreme variability exists between individuals for THC, whereas variability is much less between individuals for alcohol. These differences have not only been challenges for identifying cannabis impairment. Our success with detection of impairment for alcohol and interventions have created an illusion that cannabis-impaired driving can be addressed in the same manner as alcohol-impaired driving.

In terms of detection methods, the Breathalyzer is simple to administer, non-intrusive, provides quick results in a timely manner and BAC readings correlate closely with performance deficits. Blood tests for THC are considered the best biological specimen to assess impairment. Yet, these tests are intrusive, require training to take blood, and analyses need to be conducted in laboratory settings with possible delayed findings and have a major flaw of not being accurate measures of performance deficits at common per se levels. These impediments create insurmountable roadblocks for implementing cannabis per se laws that have comparable validity to those for alcohol.

TRAFFIC SAFETY

Two approaches for traffic safety

Undeniably, the automobile is viewed as an essential component of our lives, but produces substantial carnage with about 1.25 million traffic fatalities per year worldwide and between 20 and 50 billion people sustaining injuries (World Health Organization,

2015). The vast majority of these collisions, estimated at 94%, are due to human error (U.S. Department of Transportation, 2015). Clearly, driving can be dangerous and humans make many mistakes.

In order to address this safety issue related to driving, society has implemented two major approaches to address traffic safety. The first approach is based on ***harm reduction.*** Harm reduction is a term borrowed from drug-policy initiatives and is a set of practical strategies which are aimed at reducing the negative consequences of drug use without requiring abstinence as a goal. Interestingly, the principles of harm reduction, which have been controversial in terms of drug policy, have simply been taken for granted as logical approaches for improving safety associated with driving cars. Harm-reduction approaches aim at reducing the likelihood of collisions and injuries to drivers, while accepting the fact that people drive—an activity that is inherently dangerous. Examples of harm-reduction approaches which reduce collisions include median barriers, better lighting on roads, rumble strips, removal of roadside hazards such as trees and other solid objects, speed bumps, better and improved traffic light switching devices. These initiatives are aimed at reducing the likelihood of crashes and associated injuries by improving the environment in which people drive. Other harm-reduction initiatives are aimed at the automobile itself. Examples of these approaches include seatbelts, daytime running lights, antilock brakes, air bags, rear-view cameras and better tires. Some of these interventions, such as seatbelts and airbags, acknowledge the possibility of collisions and are aimed at reducing the extent of injuries if collisions occur, whereas other interventions, such as daytime running lights, are intended to reduce the likelihood of crashes. This harm-reduction approach is aimed at minimizing the likelihood and magnitude of potential harms from driving and accepts that driving is dangerous.

The second approach has been to provide rules of conduct for drivers to prevent crashes and facilitate the efficient movement of cars on the road. These rules (laws) require driver acceptance to be effective. The purpose behind many of these laws is self-evident to most people in that rules are needed to prevent chaos and traffic congestion. Traffic lights are a good example of rules that drivers must learn to facilitate safe driving. Other rules, such as speed limits, are less accepted. The means by which such rules are enforced is by sanctions to those who disobey them, based on the principle of deterrence. According to deterrence theory, if undesirable acts are punished, then people will be less likely to engage in those behaviours in the future (i.e. specific deterrence) and the larger population will be less likely to engage in behaviours based on fear of being punished (i.e. general deterrence). According to the theory of deterrence, if the punishment is certain, swift and severe, then people will be less likely to engage in those behaviours. For some types of offences, including impaired driving by alcohol, there is convincing empirical evidence that increased certainty of being caught has a greater deterrent effect than the severity of punishment (Wright, 2010; Nagin and Pogaesky, 2001; National Institute of Justice, 2014).

Although deterrence has benefits in controlling human behaviour, it also has drawbacks. First, deterrence is based on the premise that humans can weigh the pros and cons of any given behaviour and refrain from behaviours that are against the law. Studies show this concept is true for some people but not others. For alcohol, those with an alcohol disorder may see the benefits of driving under the influence more positively than others. This may be partially due to a high frequency of drinking and when people are under the influence of substances, rational decision-making becomes more difficult. Second, punishment or negative reinforcement is not shown to be the best way to control human behaviour, as punishment can cause unwanted side effects. Third, drivers make unintentional mistakes. For example, some drivers may not be aware they are breaking a law, a necessary condition for deterrence to be effective. These are some reasons that deterrence will not be fully effective among all people. Deterrence theory originated in the 18th century, and is of central importance to 20th century criminal justice, yet empirical evidence that it works well is weak (Paternoster, 2010). It is clear that drinking-and-driving laws have saved lives, but the costs of deterrence in terms of policing and the judiciary have been steep and the public-health issue of drinking and driving remains unacceptable.

Laurence Ross, who extensively researched the impact of drunk driving legislation worldwide, became skeptical about the feasibility of further reducing deaths by deterrent means due to the costs and will to enforce such measures over time (Ross, 2017). He recommended higher taxes on alcohol and a focus on safer roads and automobiles that would be beneficial for reducing the likelihood and severity of any type of collision. At the heart of these comments is a frustrating reality that humans make mistakes, not only by drinking and driving, but many other types of mistakes that put them at risk of collisions.

The next major level of progress will only be achieved through harm-reduction initiatives that prevent humans from making mistakes. One initiative that reduces alcohol-related casualties is the ***alcohol ignition interlock*** system, where passing a Breathalyzer test is needed to operate an automobile. The ignition interlock system when installed, shown to be effective in reducing ***recidivism*** (Elder et al., 2011), has been mandated for convicted drinking drivers.

Being born in the industrial age and entering the information age, I have seen great changes in our world. I have seen great strides in automobile safety through technology and I have often wondered about the future of transportation. We've seen a shift in major cities from automobiles to much safer public transportation. Many years back, I envisioned an "eyealyzer," a harm-reduction approach similar to alcohol ignition interlock devices, where drivers' eyes could be analysed to determine if drivers are capable of driving safely. Such an instrument could serve multiple purposes. It could be used for theft control to stop people who are unauthorized to drive a particular car. It could examine eye movements, pupil dilation and other eye characteristics for substance use and fatigue. The vehicle would not start for those who failed to pass. I also wondered if driverless cars would sometime be common, never thinking this would occur in my lifetime. Nonetheless, the concept of the driverless car is one that takes away control of potentially dangerous machinery from humans, who by our very nature are imperfect and make mistakes—albeit some make more mistakes and poorer decisions than others. Technology will eventually lead us to a safer transportation world. However, legislative approaches that address impaired driving today rely heavily on deterrence aimed at changing human behaviour.

Laws for impaired driving

Although per se laws for alcohol impairment has been defined in practically all countries of the world, this definition varies based on cultural conditions on how to strike the balance between the rights of the individuals and perceived public safety. Low BAC cut-offs of .02% alcohol in countries such as Sweden reflect a cautious approach, where it is best to err on the side of public safety versus individual rights. Although research evidence does *not* show drivers at low BAC levels of less than .04% alcohol represent a safety issue on the road, the Swedish message is not have even a single drink and drive within an hour, and to be completely sober before driving. The acceptability of such laws depends in part on cultural values and the severity of penalties to offenders. Fortunately, research has shown that high-severity penalties are **not** needed for effective deterrence. In any case, the Swedish cut-off is not so low as to prohibit drinking altogether. Those caught will have consumed alcohol within a short time frame before driving or consumed a very large amount within the past day. The message is simple to understand. Those over the limit will have increased risk of crashes compared to those below a low limit; however, many of those caught above a low limit may still not represent a real safety risk. Some may feel that we should apply similar principles to cannabis. Unfortunately, this not possible due to the pharmacokinetics of cannabis. Any low-limit amounts are tantamount to prohibition of driving among cannabis users. The time frame in which use must not occur is much longer at levels below 5 ng/mL for a THC blood test than for alcohol at .02%, perhaps up to several days for daily users, and a low limit is not simple for users to understand. Nor is it clear how much cannabis is needed to reach a certain low limit; people are not always aware of the strength of the cannabis, and also, they may share a joint with others, and therefore not know how much they used.

In North America, the legal definition of impairment for alcohol is a BAC of .08% in Canada and most U.S. states, and .05% alcohol in many Canadian provinces and some U.S. states. In Canada, the federal cut-off of .08% alcohol is consistent with severe performance deficits that virtually nearly all people would experience. The criminal code penalty for offenders is consistent with other offences, such as criminal negligence. The lower thresholds in some provinces come with no criminal record and fewer safeguards for due process.

Although definitions of alcohol impairment do differ widely around the world, any cut-off implies drinking has occurred within a relatively short time window after use. Testing for THC in blood or other biological samples is not comparable to BACs for alcohol. Alcohol is eliminated fairly quickly at a constant rate, whereas THC can remain in the body and build up for very long periods of time in daily users. This means that low threshold tests, such as those worldwide at 2 ng/mL THC or less, effectively become laws to prohibit cannabis use among drivers.

Given that research shows cannabis causes impairments, an argument could be made that in the interests of public safety we should prohibit drivers from driving even at the minimum levels at which THC can be detected. This sends the message of zero tolerance. This argument would appear to have more merit under a criminalized system than under a regime where cannabis is legal. If cannabis is illegal, then driving with traces of cannabis in one's system could also be illegal and does not necessarily need to be based on traffic-safety goals. However, under a legalized regime, if citizens are able to use cannabis in a safe manner, then they shouldn't be penalized for driving in a safe manner. Impaired driving laws within the Criminal Code of Canada must adhere to a high standard of fairness and be considered reasonable. This fairness means that laws should not be discriminatory and that individuals charged for offences should have the rights to defend themselves (i.e. due process). Laws that do not achieve this purpose undermine the system itself and create a disrespect by citizens to the justice system. Examples of this disrespect were the prohibition laws against alcohol in the early 1900s and those against cannabis making possession illegal. The truth is that research does not demonstrate impairment for the majority of individuals at levels of 5 ng/mL or below and this fact could lead to challenges against such laws.

In this book, comparisons are drawn between current legislation for impaired driving by alcohol. In particular, I compared the prevalence of use and traffic-safety harms of alcohol and cannabis, their pharmacokinetics and the scientific evidence of the validity of biological tests for detecting performance deficits. Driving under the influence of alcohol is both much more prevalent and more dangerous than cannabis while driving. The detection of impairment for alcohol is more valid than for THC. Alcohol produces more harms in terms of traffic crashes, yet proposed legislation for cannabis is more stringent. Unfortunately, research indicates cannabis cut-offs are very low, which will result in excessive proportion of false positives—drivers diagnosed as impaired who are not. The issue is greatly complicated by the mode in which people use cannabis (smoking or ingested), the length of time THC remains in the system, especially for daily users, and substantial tolerance of some users to negative performance effects of cannabis.

Canadian laws for cannabis-impaired driving

With legalization of cannabis in Canada, concerns have been raised regarding driving under the influence of cannabis. The federal government has recommended two-tiered criminal laws with a per se blood THC range of 2–5 ng/mL and a cut-off of 5 ng/mL to address cannabis-impaired driving. These laws are similar to those in many other countries in that they have legislation prohibiting driving at cut-offs of 5 ng/mL or below. I am not aware of any country that has laws against the range of 2–5 ng/mL. In order to charge a driver under the new act, the police must have reasonable suspicion to believe they are under the influence, which could include an oral fluid screening test. Drivers who test positive from point of collection on the oral fluid screening test will be subject to additional diagnostic tests aimed at determining impairment. These tests could include a DRE assessment or a blood test to detect THC concentration levels. Neither of these approaches has been properly validated although much more research has been conducted on the validity of the DEC than the validity of blood tests. DRE laws have been criticized for being cumbersome requiring costly training of officers, subject to errors and can be challenged in court. However, the consecutive stages of assessment reduce the likelihood of false positives (Gordis, 2014). The per se blood limit recently passed by the Canadian government offers a more streamlined approach but may be subject to even greater inaccuracy.

A shortcoming of the new Canadian legislation is the use of oral fluid tests. The Drager 5000 point-of-collection oral fluid test is being authorized for police to use as a screening test. The cut-off for the test will be 25 ng/mL. This number sounds high but it is not for two major reasons. First, the test is based on an immunoassay procedure. Recall that screening test of this type cannot distinguish THC from other compounds of a similar molecular structure, such as carboxy-THC or hydroxy-THC. This means that the cut-off of 25 ng/mL includes a combination of these different compounds, not simply THC. The second reason it is low is that concentration levels of THC are much greater in oral fluids than in blood—estimated between 20 to 60 times greater. Taking the lower ratio, this means 20 ng/mL of THC in oral fluid would be equivalent to 1 ng/mL in blood. In actual fact, this cut-off is very low compared to a blood sample. One might argue that it doesn't matter because these tests can't be used to convict a driver; however, a positive test will likely create considerable anxiety for a driver. The oral fluid test cut-off is much less stringent than the blood test and speaks to the issue of fairness in the criminal justice system. In addition, there are other reasons to question the use of these oral fluid tests in Canada. Oral fluid tests were pilot tested the under Canadian conditions by police, and 65% of the tests were administered outside the manufacturer's suggested operating temperatures. The validity of the tests is unknown at these temperatures. Several police forces, including the Vancouver police, have already stated they will not be using the tests for these and other reasons (CBC News, 2018).

Another limitation with the Canadian legislation is that for THC in the range of 2–5 ng/mL in whole blood, existing research suggests most drivers don't represent a safety risk. In fact, a high-quality observational study indicates drivers testing positive below 5 ng/mL are protective of being responsible for fatal crashes (secondary analysis of Drummer et al., 2004, reported in Grotenhermen et al., 2007). Findings from this field study is supported by studies that show cannabis is a risk-aversive drug. In other words, those who are feeling the effects of cannabis may compensate in ways that make their driving safer. This could include actions such as driving more defensively by slowing down and leaving more distance when driving behind a car. Research indicates occasional users are more adversely affected by cannabis, a group that is less likely to drive after use than regular users. Laboratory studies have also not shown drivers in this range represent a meaningful safety risk. This choice of a low range with no scientific evidence of crash risk is less justifiable than a low cut-off. A range of 2–5 ng/mL THC in blood cannot be considered evidence-based.

Higher cut-offs will be related to higher likelihood of impairment. Data from the Drummer et al. (2004) study (reported in Grotenhermen et al., 2007) is suggestive that a cut-off in the range of 6–8 ng/mL is equivalent to about .05% alcohol and Hartman et al (2015b) suggested that 8 ng/mL is equivalent to .05% alcohol. These two empirical studies provide some evidence that a range of 6–8 ng/mL could be used as a conservative and tentative impairment threshold; however, even this conclusion is highly tentative based on sparse research.

The Canadian laws for cannabis impaired driving are more streamlined than the DRE process but not perfect. The procedures to obtain blood and then to analyse it have not been well specified. Since THC is not eliminated in blood at a constant rate, blood samples must be obtained at the time of suspected impairment. Also, since confirmatory tests using chromatography and spectrometry methods are needed for accurate assessment of THC concentrations but require time consuming laboratory methods, it is not clear how analyses will be carried out in a timely manner. Most of the conclusions made in this book were known when the DRE process was developed; and the DRE methods were proposed with acknowledgement of their limitations. A major issue is that the validity of blood tests at the levels proposed have not been scientifically proven.

Although the DRE approach has limitations, it has major strengths against the Canadian legislation for driving while impaired by cannabis. The main strengths is that it focusses more on behavioural criteria related to being under the influence of cannabis rather than untested and dubious criteria based on low blood THC thresholds. Clearly, legislation that makes our roads safer is a good thing. However, the rationale behind penalizing drivers within a low range of 2-5 ng/mL THC is dubious, as this range has not been clearly demonstrated in any laboratory study as meaningfully related to crash risk and has actually been shown to be a protective factor in a frequently cited observational study of culpability risk. Papers that support per se legislation at these low thresholds are not rooted in traditional principles of scientific inquiry. It is inconceivable from an evidence-based perspective that such legislation will make our roads measurably safer. Legislation that has little scientific basis to improve traffic safety has the potential to do more harm than good.

Higher THC levels in blood might be used to draw reasonably valid conclusions of impairment for those who smoked cannabis, but more research is needed. Research suggests an average individual who tests positive at 10 to 20 ng/mL THC would have likely smoked cannabis within the hour, which might be a suitable measure of impairment. Legislation could be passed prohibiting smoking of cannabis in cars. However, if cannabis was ingested or if the individual tested was a daily user, this cut-off may not meet common scientific standards of validity. As well, there is suggestive evidence that testing for both THC and hydroxy-THC may provide a more valid indication of impairment.

Implications of the Canadian laws on various stakeholders

The material in this book has implications for different groups, and below are suggestions on some issues of importance for cannabis impairment detection.

Federal government of Canada—My sense is that a legal challenge will be coming, especially in relation to the per se law forbidding drivers with a THC content between 2-5 ng/mL. In the review, not a single credible study was found that shows most drivers in this range represent a safety risk for driving. It would be good for the government to refine the rationale for the policy that is consistent with prevailing research, but this may be impossible. The THC blood cut-offs on cannabis and driving are very close to a zero-tolerance approach and will lead to unfairness within the Criminal Code of Canada.

Provincial governments in Canada—These governments have much more leeway in terms of implementing traffic-safety laws without court challenges. Driving is a privilege, not a right. Provincial governments could consider roadside oral fluid testing of drivers with a mobile laboratory that can conduct confirmatory "gold standard'" drug tests, specifically for THC. For such legislation, research still is needed to assess whether a valid cut-off can be established. Other types of legislation focusing on behavioural measures, such as the finger-to-nose and modified Romberg balance test, indicators of divided attention, could be considered. These divided-attention tests appear to be fairer criteria than drug tests to classify individuals as roadworthy. If drivers cannot pass these tests for any reason, they may be risky drivers as well, and subject to provincial penalties including forfeiture of driving privileges. If a gold standard of cannabis impairment is to be achieved, it will be through the development of indicators of direct performance measures, and ideally, performance symptoms that are specific to cannabis impairment.

Lawyers—These laws under the federal government are an opportunity to challenge. A challenge likely won't be popular, but the world will be better with laws based on evidence-based policies.

Users of cannabis—It is not a good idea to drive when feeling high. Also, for users who do not wish to have a criminal record but drive a car, I suggest leaving your "Stoned Again" T-shirt at home (available online in several designs and colours from several distributors). The shirt could create suspicion of cannabis use and given the long detection window at low cut-offs, a positive test could occur. Also, brush one's teeth before driving is good practice after use, as an oral fluid test could pick up residues of smoking cannabis and trigger a positive test. Although ingesting cannabis rather than smoking will produce lower blood THC levels, I do not recommend this as a strategy to pass a blood test. Eating pot produces a more intense high than smoking and for a longer period of time, making the risks of crashes unacceptable.

Police—I have learned that police are not appreciative of advice from others, so in this case I will describe a likely scenario. The police have discretion to enforce laws, which could lead to a very uneven application of these laws on Canadian drivers. In fact, based on my recent study in B.C. from the period before legalization, about 95% of people found in possession of cannabis were not charged by police (Macdonald et al., 2018). Some officers will rarely order blood tests, seeing the laws as unfair and overreaching in that many of those caught are actually safe to drive. Others may choose to use these laws selectively on certain segments of society they see as undesirable. Still others will vigilantly uphold the laws as much as possible. These scenarios leads to a patchwork of enforcement that is uneven.

Laws that are seen as unfair can be undermined by those who oppose them. Police in Canada have used police discretion to practically ignore cannabis possession laws. The failure of the criminal justice system to uphold these laws is likely an important factor leading to legalization in Canada. Similarly, per se laws for cannabis, where the threshold is so low that the majority of drivers at this level do not represent a significant crash threat, may also be rarely used by police.

LIMITATIONS AND DIRECTIONS FOR FUTURE RESEARCH

A major theme throughout this book is greater adoption of the epidemiological method for assessing the validity of biological tests for impairment. The acute effects of cannabis do cause performance deficits that can meet standards of impairment, but how to measure them has been elusive.

One issue is "the elephant in the room," a major issue of importance that is rarely discussed: *How do we define impairment*? Impairment is essentially a legal definition where performance deficits are considered severe enough to warrant sanctions. More research is needed on defining the magnitude of performance deficits in laboratory conditions that constitute impairment. As well, more research should go into establishing the criteria for impaired performance after taking cannabis, building on progress by the working group led by Kay and Logan (2011). This work appears under-recognized and needs to be more prominent in moving science forward. Another approach to defining impairment is through drawing on the studies of alcohol deficits at different BACs and using these performance deficits to calibrate equivalent thresholds for cannabis similar to the approach used by Hartman et al. (2015b).

Once these performance criteria have been clearly defined, experimental laboratory studies can be conducted to determine optimal biological cut-offs that best detect these impairment thresholds. To achieve this goal, the validity of the tests for specific thresholds should be assessed using the epidemiological methods described in Chapter 2. Although people will disagree on the optimal trade-off between sensitivity and specificity, major disagreements will only occur in situations where the trade-off is extreme. Presently, the state of our knowledge on these issues for any cut-off remains poor, and cut-offs are guesswork based on the wrong kind of studies. Only with this knowledge on the validity of drug tests can legislators adopt evidence-based policies and provide a rationale for them.

One issue that creates challenges for interpreting, comparing and integrating findings is the diverse ways in which cannabis research is carried out. Although some diversity in research designs is good for science, lack of standardization of how researchers' conceptualized importance constructs is a particular problem in cannabis research. Greater standardization is recommended in relation to several areas of cannabis research as follows:

1. Researchers should quantify the amount of cannabis administered more consistently in laboratory studies. A particular concern was studies that only reported the THC content but not the weight of the product, which meant that the actual THC consumed was unknown. The best way to define the amount consumed is by expressing the dosage in terms of mg of THC.
2. Another issue that made it difficult to compare findings of studies is the diversity in which blood was analysed as either whole blood, serum or plasma. As pointed out in this book, although some researchers claim near constant serum/plasma-to-whole-blood THC ratios that support equivalency, research shows the ratio between the two is not sufficiently constant between people to adopt a universal conversion ratio. THC concentration levels in serum and plasma can be considered equivalent. Overall, the use of whole blood appears more common for legislative limits and in research, and appears to be the best medium to analyse THC in blood.
3. Some studies reported use of screening drug tests whereas others used more accurate confirmation drug tests with chromatographic and spectrometry methods. All studies should include confirmation methods, unless a specific rationale is given, such as convenience of administration, and inaccuracies associated with screening tests are stated as a limitation.

In addition to the above, there should be universal acceptance that urine tests should not be used to detect impairment. Although this conclusion is well accepted in the traffic-safety world, it is still contentious among some in Canada who support urinalysis for employees. Five observational studies were reviewed that found no significant relationship between positive urine tests for cannabinoids and crashes. As well, many studies were subject to selection bias due to consent being required to obtain biological specimens from control subjects but not cases. This type of bias increases the likelihood of significant yet erroneous findings.

More research is needed on the validity of oral fluid *confirmation* tests. A reasonable cut-off for use within the past hour would need to be established, which may not be feasible. Since such tests are currently being used in countries such as Australia, validity studies to determine cut-offs associated with impairment are needed. Perhaps mobile laboratories, where confirmatory tests could be conducted in a timely manner may be worthwhile for more consideration as a deterrence-based intervention approach. Currently, laws that penalize drivers based on oral fluid confirmation tests are not evidence-based.

Another challenge is to better understand the role of hydroxy-THC in performance deficits. Hydroxy-THC is a psychoactive metabolite created by the body's breakdown of THC that is redistributed in the bloodstream. We need to learn more about the differences between smoked and ingested cannabis in terms of performance deficits. Comparisons of studies by Hunault et al. (2014), where subjects smoked cannabis, and Ménétrey et al. (2005), where subject ingested cannabis, shows large differences in performance effects, their durations, and concentration levels of THC in the blood. More research is needed on two major variables related to detection periods and cut-offs for smoked versus ingested cannabis.

Observational studies that used biological tests (whether blood, oral fluids or urine) to assess the relationship between crashes and cannabis typically chose extremely low cut-offs of compounds such as THC. This approach is surprising, given that drug concentrations can be measured as a continuous variable and choosing to divide the data into two arbitrary groups reduces the likelihood of finding statistical significance. The strength of the relationships (i.e. effect sizes) were very low in most studies, typically under 2 and many studies failed to find statistical significance. If the drug concentrations are analysed as a continuous variable in relation to crashes and there is a dose-response relationship between concentration levels and crashes, the likelihood of statistical significance is increased because this measure is more precise. As pointed out by White (2017), dose-response relationships have not been convincingly shown in research. However, given the preponderance of the evidence for the acute effects of cannabis, greater doses of cannabis and ingestion as opposed to smoking are related to increased performance deficits.

In this book, I have compared our knowledge of alcohol and cannabis in relation to driving and Canadian legislation to reduce driving casualties. The safety benefits of the Breathalyzer for detecting alcohol impairment are substantial and undeniable. A large body of credible research shows alcohol negatively affects performance in a dose-response relationship and the BAC level from breath tests is an excellent measure of performance deficits. As well, alcohol is eliminated from the body at a constant rate and is water-soluble, which means there is a close relationship between alcohol in the blood and performance. Although alcohol can cause long-term effects, such as those due to hangovers, the

Breathalyzer cannot detect such deficits. For cannabis detection, the story is much different. The blood tests are inconvenient, difficult to administer in a timely fashion and have poor relationships between low levels of THC and performance deficits. Based on the preponderance of research, a range between 2–5 and a cut-off above 5 ng/mL for impairment are very low compared to the alcohol cut-offs at .05%. As well, the theory of 24-hour impairment does not provide an evidence-based rationale for adopting low THC cut-off levels. The preponderance of research evidence shows cannabis hangovers are not meaningfully related to performance deficits. A very conservative lower per se impairment cut-off is 6 ng/mL and can be justified based on one empirical observational study, and 8 ng/mL based on one laboratory study as equivalent to .05% alcohol. These studies have not been replicated or validated. No credible evidence of meaningful impairment below 5 ng/mL exists for smoked cannabis.

Bibliography

Adams, I. B., & Martin, B. R. (1996). Cannabis: pharmacology and toxicology in animals and humans. *Addiction, 91*(11), 1585-1614.

Alcohol Help Center (undated) Blood alcohol calculator. Retrieved from http://www.alcoholhelpcenter.net/Program/BAC_Standalone.aspx

Allen, J.P., Litten, R.Z., Strid, N., & Sillanaukee, P. (2001). The role of biomarkers in alcoholism medication trials. *Alcoholism: Clinical and Experimental Research, 25*(8), 1119–1125.

Allsop, D. J., Norberg, M. M., Copeland, J., Fu, S., & Budney, A. J. (2011). The Cannabis Withdrawal Scale development: patterns and predictors of cannabis withdrawal and distress. *Drug and Alcohol Dependence, 119*(1–2),123–129.

AMA Council on scientific affairs. (1986). Alcohol and the driver. *JAMA, 255(*4), 522-527.

Ames, G. M., Grube, J. W., & Moore, R. S. (1997). The relationship of drinking and hangovers to workplace problems: an empirical study. *Journal of Studies on Alcohol, 58*(1), 37-47.

Andås, H. T., Krabseth, H.-M., Enger, A., Marcussen, B. N., Haneborg, A.-M., Christophersen, A. S., ... Øiestad, E. L. (2014). Detection Time for THC in Oral Fluid After Frequent Cannabis Smoking. Therapeutic Drug Monitoring, 36(6), 808–814.

Andréasson, R., & Jones, A. W. (1995). Erik M.P. Widmark (1889-1945): Swedish pioneer in forensic alcohol toxicology. *Forensic Science International, 72*(1), 1–14.

Arendt, M., Munk-Jørgensen, P., Sher, L., & Jensen, S.O.W. (2013). Mortality following treatment for cannabis use disorders: predictors and causes. *Journal of Substance Abuse Treatment, 44*(4), 400–6.

Asbridge, M., Hayden, J. A., & Cartwright, J. L. (2012). Acute cannabis consumption and motor vehicle collision risk: systematic review of observational studies and meta-analysis. *British Medical Journal*, e536–e536.

Asbridge, M., Mann, R., Cusimano, M. D., Trayling, C., Roerecke, M., Tallon, J. M., Whipp, A., & Rehm, J. (2014). Cannabis and traffic collision risk: findings from a case-crossover study of injured drivers presenting to emergency departments. *International Journal of Public Health, 59*(2), 395–404.

Ashton, C. H. (1999). Adverse effects of cannabis and cannabinoids. *British Journal of Anaesthesia, 83*(4), 637–49.

Ashton, C. H. (2001). Pharmacology and effects of cannabis: a brief review. *The British Journal of Psychiatry, 178*(2), 101–106.

Atakan, Z. (2012). Cannabis, a complex plant: different compounds and different effects on individuals. *Therapeutic Advances in Psychopharmacology, 2*(6), 241–54.

Babor, T. F., Caetano, R., Casswell, S., Edwards, G., Giesbrecht, N., Graham, K., and others. (2010). *Alcohol: No ordinary commodity: Research and public policy*. (2nd ed.). Oxford University Press.

Bates, M. N., & Blakely, T. A. (1999). Role of cannabis in motor vehicle crashes. *Epidemiologic Reviews, 21*(2), 222–232.

Beasley, E., Beirness, D., & Porath-Waller, A. (2011). *A comparison of drug and alcohol-involved motor vehicle driver fatalities*. Ottawa, ON: Canadian Centre on Substance Abuse.

Beccaria C. (1764). *On crimes and punishment*. New York, New York: MacMillan.

Beirness, D.J., LeCavalier, J., & Singhal, D. (2007). Evaluation of the drug evaluation and classification program: a critical review of the evidence. *Traffic Injury Prevention, 8*, 368-376.

Beirness, D.J. & Davis, C.G. (2007). Driving after drinking in Canada: findings from the Canadian Addiction Survey. *Canadian Journal of Public Health, 98*(6), 476–480.

Beirness, D.J., Beasley, E., & Lecavalier, J. (2009). The accuracy of evaluations by drug recognition experts in Canada. *Journal of the Canadian Society of Forensic Science*, 42(1), 75-79.

Beirness, D. J., & Beasley, E. E. (2010). A roadside survey of alcohol and drug use among drivers in British Columbia. *Traffic Injury Prevention, 11*(3), 215–21.

Beirness, D. J., Beasley, E. E., & Boase, P. (2013). Drug use among fatally injured drivers in Canada. In B. Watson & M. Sheehan (Eds.), *Proceedings of the 20th International Conference on Alcohol, Drugs and Traffic Safety*. Brisbane: Centre for Accident Research and Road Safety.

Beirness, D. J., & Smith, D. A. R. (2017). An assessment of oral fluid drug screening devices. Canadian Society of Forensic Science Journal, 50(2), 55-63.

Berghaus, G., & Guo, B.L. (1995). Medicines and driver fitness - findings from a meta-analysis of experimental studies as basic information to patients, physicians and experts. In C.N Kloeden & A.J. McLean (Eds.). *Proceedings of the 13th International Conference on Alcohol, Drugs and Traffic Safety* (pp. 295-300). Adelaide, Australia: University of Adelaide.

Berghaus, G., Sheer, N., & Schmidt, P. (1995). Effects of cannabis on psychomotor skills and driving performance - a meta-analysis of experimental studies. In *Proceedings of the 13th International Conference on Alcohol Drugs and Traffic Safety* (pp. 403-409). Adelaide, Australia: The University of Adelaide, NHMRC Road Accident Research Unit.

Biecheler, M.B., Peytavin, J.F., Facy, F., & Martineau, H. (2008). SAM survey on "drugs and fatal accidents": search of substances consumed and comparison between drivers involved under the influence of alcohol or cannabis. *Traffic Injury Prevention,* 9(1), 11–21.

Blencowe, T., Pehrsson, A., Lillsunde, P., Vimpari, K., Houwing, S., Smink, B., Mathijssen, R., Van der Linden, T., Legrand, S., Pil, K., & Verstraete, A. (2011). An analytical evaluation of eight on-site oral fluid drug screening devices using laboratory confirmation results from oral fluid. *Forensic Science International, 208,* 173-179.

Blomberg, R. D., Peck, R. C., Moskowitz, H., Burns, M., & Fiorentino, D. (2005). *Crash risk of alcohol involved driving: A case-control study*. Washington: DC, National Highway Traffic Safety Administration.

Blomberg, R. D., Peck, R. C., Moskowitz, H., Burns, M., & Fiorentino, D. (2009). The Long Beach/Fort Lauderdale relative risk study. *Journal of Safety Research, 40*(4), 285–92.

Blows, S., Ivers, R. Q., Connor, J., Ameratunga, S., Woodward, M., & Norton, R. (2005). Marijuana use and car crash injury. *Addiction, 100*(5), 605–11.

Bogstrand, ST., Gjerde, H., Normann, PT., Rossow, I., & Ekeberg, Ø. (2012). Alcohol, psychoactive substances and non-fatal road traffic accidents--a case-control study. *BMC Public Health, 12*(1), 734.

Bondallaz, P., Favrat, B., Chtioui, H., Fornari, E., Maeder, P., & Giroud, C. (2016). Cannabis and its effects on driving skills. *Forensic Science International, 268*, 92–102.

Borkenstein, R. F., Crowther, R. F., Shumate, R. P., Ziel, W. B., & Zylman, R. (1964). *The Role of the Drinking Driver in Traffic Accidents.* Bloomington: IN, Department of Police Administration, Indiana University.

Borkenstein, R. F. (1985). Historical perspective: North American traditional and experimental response. *Journal of Studies on Alcohol,* (10), 3-12.

Bosker, W.M., & Huestis, M.A. (2009). Oral fluid testing for drugs of abuse. *Clinical Chemistry, 55*(11), 1910–31.

Boy, R. G., Henseler, J., Ramaekers, J. G., Mattern, R., & Skopp, G. (2009). A comparison between experimental and authentic blood/serum ratios of 3,4-methylenedioxy-methamphetamine and 3, 4-methylenedioxyamphetamine. *Journal of analytical toxicology, 33*(5), 283-286.

Brady, J.E., Baker, S. P., Dimaggio, C., McCarthy, M. L., Rebok, G. W., & Li, G. (2009). Effectiveness of mandatory alcohol testing programs in reducing alcohol involvement in fatal motor carrier crashes. *American Journal of Epidemiology, 170*(6), 775–82.

Brady, J.E., & Li, G. (2013). Prevalence of alcohol and other drugs in fatally injured drivers. *Addiction, 108*(1), 104–14.

Brault, M., Dussault, C., Bouchard, J., & Lemire, A. M. (2004,August). The contribution of alcohol and other drugs among fatally injured drivers in Quebec: final results. In *Proceedings of the 17th International Conference on Alcohol, Drugs and Traffic Safety*. Glasgow, Scotland.

Brenneisen, R., Meyer, P., Chtioui, H., Saugy, M., & Kamber, M. (2010). Plasma and urine profiles of Delta9-tetrahydrocannabinol and its metabolites 11-hydroxy-Delta9-tetrahydrocannabinol and 11-nor-9-carboxy-Delta9-tetrahydrocannabinol after cannabis smoking by male volunteers to estimate recent consumption by athletes. *Analytical and Bioanalytical Chemistry, 396*(7), 2493–2502.

Brown, S., Vanlaar, W., Mayhew, D. (2013). *The Alcohol-Crash Problem in Canada: 2010.* Ottawa: Traffic Injury Research Foundation.

Broughton, M. (2014). The Prohibition of Marijuana. *Manitoba Policy Perspectives, 1*(1), 1–4. Retrieved from http://staff.uob.edu.bh/files/620922311_files/Prohibition-of-Gharar.pdf

Broyd, S. J., Van Hell, H. H., Beale, C., Yücel, M., & Solowij, N. (2016). Acute and chronic effects of cannabinoids on human cognition - A systematic review. *Biological Psychiatry, 79*(7), 557–567.

Brubacher, J.R., Chan, H., Martz, W., Schreiber, W., Asbridge, M., Eppler, J., Lund, A., Macdonald, S., Drummer, O., Pursell, R., Andolfatto, G., Mann, R., & Brant, R. (2016). Prevalence of alcohol and drug use in injured British Columbia drivers. *BMJ Open, 6*(3), 1-10.

Budney, A J., Hughes, J. R., Moore, B. a, & Novy, P. L. (2001). Marijuana abstinence effects in marijuana smokers maintained in their home environment. *Archives of General Psychiatry, 58*(10), 917–924.

Burns, M., & Anderson, E. W. (1995). A Colorado Validation Study of the Standardized Field Sobriety Test (SFST) Battery. National Highway Traffic Safety Administration, Office of Transportation Safety, Colorado Department of Transportation (95-408-17-05) (95).

Butler, B. & Tranter, D. (1994) Behavioural tests to assess performance. In S. Macdonald, & P. Roman (Eds.). *Research Advances in Alcohol and Drug Problems.* New York: Plenum Press.

Byas-Smith, M. G., Chapman, S. L., Reed, B., & Cotsonis, G. (2005). The effect of opioids on driving and psychomotor performance in patients with chronic pain. *Clinical Journal of Pain, 21*(4), 345–352.

Callaghan, R. C., Gatley, J. M., Veldhuizen, S., Lev-Ran, S., Mann, R., & Asbridge, M. (2013). Alcohol or drug-use disorders and motor vehicle accident mortality: a retrospective cohort study. *Accident Analysis & Prevention, 53,* 149–55.

Campbell, D. & Stanley, J. (1963). *Experimental and quasi-experimental designs for research.* Chicago, IL: Rand-McNally.

Cannabix technologies. Retrieved from www.cannabixtechnologies.com/thc-breathalyzer.html

Canadian Alcohol and Drug Use Monitoring Surveys (2004-2011). Retrieved from: http://www.hc-sc.gc.ca/hc-ps/drugs-drogues/stat/ 2011/summary-sommaire-eng.php

Canadian Public Health Association. (2006). *Drinking and Alcohol* [Fact Sheet]. Retrieved from http://www.cpha.ca/uploads/progs/ /drinkingfacts/facts e.pdf

Capler, R., Bilsker, D., Van Pelt, K., & MacPherson, D. (2017). Cannabis use and driving: Evidence Review. Retrieved from https://drugpolicy.ca/wp-content/uploads/2017/02/CDPC Cannabis-and-Driving Evidence-Review FINALV2 March27-2017.pdf

Cary, P. (2004). Urine drug concentrations: The scientific rationale for eliminating the use of drug test levels in court proceedings. *National Drug Court Institute, 4*(1), 1–8.

Cashman, C.M., Ruotsalainen, J.H., Greiner, B.A., Beirne, P.V., & Verbeek, J. (2009). Alcohol and drug screening of occupational drivers for preventing injury. *Cochrane Database of Systematic Reviews, 2009*(2), 1–23.

CBC News (2018) Pair of police forces in B.C. decides against using federally approved pot-screening device. (September 25). Retrieved from

https://www.cbc.ca/news/canada/british-columbia/vancouver-police-delta-police-drager-marijuana-testing-device-1.4837466
Centers for Disease Control and Prevention (2012). Motor vehicle crash deaths in Metropolitan Areas — United States, 2009. *Morbidity and Mortality Weekly Report, 61*(28), 523-528.
Centers for Disease Control and Prevention (2013). *Web-based Injury Statistics Query and Reporting System (WISQARS)* [Online]. Retrieved from: http://www.cdc.gov/injury/wisqars/
Centers for Disease Control and Prevention (2016). *Impaired Driving* [Fact Sheet].Retrieved from https://www.cdc.gov/motorvehiclesafety/impaired_driving/impaired-drv_factsheet.html
Centers for Disease Control and Prevention. (2017). *What Are the Risk Factors for Lung Cancer?* [Fact sheet]. Retrieved May 21, 2018, from https://www.cdc.gov/cancer/lung/basic_info/risk_factors.htm
Chait, L. (1990). Subjective and behavioral effects of marijuana the morning after smoking. *Psychopharmacology, 100*(3), 328–333.
Chamberlain, E. & Solomon, R. (2010). Enforcing impaired driving laws against hospitalized drivers: the intersection of healthcare, patient confidentiality, and law enforcement. *Windsor Rev. Legal & Soc. Issues, 29*(1), 45–88.
Clarke, J., & Wilson, J.F. (2005). Proficiency testing (external quality assessment) of drug detection in oral fluid. *Forensic Science International, 150* (2), 161-164.
Coambs, R.B., & McAndrews, M.P. (1994). The effects of psychoactive substances on workplace performance. In: Macdonald, S. & Roman, P. (Eds.), *Drug testing in the workplace*. New York: Plenum Press.
Committee on the Health Effects of Marijuana: An Evidence Review and Research Agenda (2017). Chapter 15, Challenges and Barriers in Conducting Cannabis Research. The National Academies Press, Washington, D.C.
Compton, R. P., & Berning, A. (2015). *Research note drug and alcohol crash risk. Behavioral safety research.* Washington, DC. Retrieved from https://www.nhtsa.gov/behavioral-research/drug-and-alcohol-crash-risk-study
Compton, R.P., Blomberg, R.D., Moskowitz, H., Burns, M., Peck, R.C. & Fiorentino, D. (2002). Crash risk of alcohol impaired driving. *Proceedings of the 16th International Conference on alcohol, drugs and traffic safety (CD-ROM)*. Montreal, Canada.
Compton, R., Vegega, M., & Smither, D. (2009). *Drug-Impaired Driving: Understanding the Problem and Ways to reduce it: A report to Congress*. Washington, DC.
Cone, E., & Huestis, M. (2007) Interpretation of oral fluid tests for drugs of abuse. *Annals of the New York Academy of Science, 1098*(1), 51-103.
Cone, E., Johnson, R.E., Darwin, W.D., Yousefnejad, D., Mell, L.D., & Paul, B.D. (1987). Passive inhalation of marijuana smoke: urinalysis and room air level of delta-9 tetrahydrocannabinol. *Journal of Analytical Toxicology, 11*, 89-96.
Cone, E.J. (2001). Legal, workplace, and treatment drug testing with alternate biological matrices on a global scale. *Forensic Science International, 121*, 7-15.

Cone, E.J., Depriest, A.Z., Heltsley, R., Black, D.L., Mitchell, J.M., Lodico, C., & Flegel, R. (2015). Prescription Opiods. III. Disposition of Oxycodone in Oral Fluid and Blood Following Controlled Single-Dose Administration. *Journal of Analytical Toxicology, 39*(3), 192–202.

Cook, P.J. (1980). Research in Criminal Deterrence: Laying the Groundwork for the Second Decade. *Crime and Justice (2)*, 211-68.

Couper, F., & Logan, B. (2004). *Drugs and human performance* [Fact Sheet]. Washington: DC. Retrieved from https://www.wsp.wa.gov/breathtest/docs/webdms/DRE_Forms/Publications/drug/Human_Performance_Drug_Fact_Sheets-NHTSA.pdf

Crean, RD; Crane, NA; Mason, B. (2011). An Evidence Based Review of Acute and Long-Term Effects of Cannabis Use on Executive Cognitive Functions. *Journal of Addiction Medicine*, 5(1), 1–8.

Curran, H. V., Brignell, C., Fletcher, S., Middleton, P., & Henry, J. (2002). Cognitive and subjective dose-response effects of acute oral D 9 -tetrahydrocannabinol (THC) in infrequent cannabis users. *Psychopharmacology*, 61–70.

Dawson, D., & Reid, K. (1997). Fatigue, alcohol and performance impairment. *Nature, 388*(6639), 235.

Declues, K., Perez, S., & Figueroa, A. (2016). A Two-Year Study of Δ 9 Tetrahydrocannabinol Concentrations in Drivers; Part 2: Physiological Signs on Drug Recognition Expert (DRE) and non-DRE Examinations. *Journal of Forensic Sciences*, 63(2), 583–587.

Del Rio, M. C., & Álvarez, F. J. (1995). Illegal drugs and driving. *Journal of traffic medicine, 23*(1), 1-5.

Desrosiers, N.A., Lee, D., Concheiro-Guisan, M., Scheidweiler, K.B., Gorelick, D. A., & Huestis, M.A. (2014). Urinary cannabinoid disposition in occasional and frequent smokers: is THC-glucuronide in sequential urine samples a marker of recent use in frequent smokers? *Clinical Chemistry, 60*(2), 361–72.

Desrosiers, N. A., Himes, S. K., Scheidweiler, K. B., Concheiro-Guisan, M., Gorelick, D. A., & Huestis, M. A. (2014). Phase I and II cannabinoid disposition in blood and plasma of occasional and frequent smokers following controlled smoked cannabis. *Clinical Chemistry, 60*(4), 631–643.

Dickson, L. (2018) *Breath test can't be refused under new drunk-driving law*. Times Colonist. (July 3, 2018), D8.

Donelson, A. C. & Beirness, D.J. (1985). *Legislative issues related to drinking and driving.* Ottawa: Dep. of Justice, Policy, Programs and Research Branch.

Drummer, O. H., Gerostamoulos, J., Batziris, H., Chu, M., Caplehorn, J., Robertson, M. D., & Swann, P. (2004). The involvement of drugs in drivers of motor vehicles killed in Australian road traffic crashes. *Accident Analysis & Prevention, 36*(2), 239–248.

Drummer, O.H. (2006). Drug testing in oral fluid. *The Clinical Biochemist Reviews, 27*(3), 147-159.

Drummer, O.H., Gerostamoulos, J., Batziris, H., Chu, M., Caplehorn, J., Robertson, M.D., & Swann, P. (2004). The involvement of drugs in drivers of motor vehicles killed in Australian road traffic crashes. *Accident Analysis & Prevention, 36*(2), 239–248.

Dussault, C., Brault, M., Bouchard, J., & Lemire, A. M. (2002). The Contribution of Alcohol and Other Drugs Among Fatally Injured Drivers in Quebec: Some Preliminary Results. In Road Traffic and Psychoactive Substances (pp. 1–9). Strasbourg, France. Retrieved from http://trid.trb.org/view.aspx?id=746344

Dyer, K.R., & Wilkinson, C. (2008). The detection of illicit drugs in oral fluid: another potential strategy to reduce illicit drug-related harm. *Drug and Alcohol Review, 27*, 99-107.

Elder, R. W., Shults, R. A., Sleet, D. A., Nichols, J. L., Zaza, S., & Thompson, R. S. (2002). Effectiveness of Sobriety Checkpoints for Reducing Alcohol-Involved Crashes. *Traffic Injury Prevention*, 3(4), 266–274.

Elder, R. W., Voas, R., Beirness, D., Shults, R. a, Sleet, D. a, Nichols, J. L., & Compton, R. (2011). Effectiveness of ignition interlocks for preventing alcohol-impaired driving and alcohol-related crashes: a Community Guide systematic review. *American Journal of Preventive Medicine*, *40*(3), 362–376.

Els, C., Amin, A., & Straube, S. (2016). Marijuana and the Workplace. *Canadian Journal of Addiction, 4*, 5–7.

Elvik, R. (2013). Risk of road accident associated with the use of drugs: a systematic review and meta-analysis of evidence from epidemiological studies. *Accident Analysis & Prevention, 60*, 254-267.

European Monitoring Centre for Drugs and Drug Addiction, (2012) Driving Under the Influence of Drugs, Alcohol and Medicines in Europe — findings from the DRUID project. Luxembourg: Publications Office of the European Union

Fabritius, M., Chtioui, H., Battistella, G., Annoni, J. M., Dao, K., Favrat, B., & Giroud, C. (2013). Comparison of cannabinoid concentrations in oral fluid and whole blood between occasional and regular cannabis smokers prior to and after smoking a cannabis joint. *Analytical and Bioanalytical Chemistry, 405*, 9791–9803.

Fant, R.V, Heishman, S.J., Bunker, EB., & Pickworth, W.B. (1998). Acute and residual effects of marijuana in humans. *Pharmacology, Biochemistry, and Behavior, 60*(4), 777–84.

Fell, J., & Voas, R. B. (2014). The effectiveness of a 0.05 blood alcohol concentration (BAC) limit for driving in the United States. *Addiction, 109*(6), 869–874.

Fergusson, D. M., & Horwood, L. J. (2001). Cannabis use and traffic accidents in a birth cohort of young adults. *Accident Analysis & Prevention*, 33(6), 703–11.

Ferrara, S.D., Giorgetti, R., & Zancaner, S. (1994). Psychoactive substances and driving: State of the art and methodology. *Alcohol, Drugs and Driving, 10*, 1-55.

Fierro, I., González-Luque, J.C., & Álvarez, F.J. (2014). The relationship between observed signs of impairment and THC concentration in oral fluid. *Drug and Alcohol Dependence*, 144, 231–238.

Finnigan, F., Hammersley, R., & Cooper, T. (1998). An examination of next - day hangover effects after a 100 mg/100 mL dose of alcohol in heavy social drinkers. *Addiction, 93*(12), 1829-1838.

Gainsford, A.R., Fernando, D.M., Lea, RA, & Stowell, A.R. (2006). A large-scale study of the relationship between blood and breath alcohol concentrations in New Zealand drinking drivers. *Journal of Forensic Sciences, 51*(1), 173–8.

Gjerde, H., & Verstraete, A. G. (2011). Estimating equivalent cut-off thresholds for drugs in blood and oral fluid using prevalence regression: A study of tetrahydrocannabinol and amphetamine. *Forensic science international, 212*(1-3), 26-30.

Gjerde, H., Christophersen, A.S., Normann, P.T., & Mørland, J. (2013). Associations between substance use among car and van drivers in Norway and fatal injury in road traffic accidents: A case-control study. *Transportation Research Part F: Traffic Psychology and Behaviour*, 17, 134–144.

Gjerde, H., Strand, M. C., & Mørland, J. (2015). Driving under the influence of non-alcohol drugs — an update part i: epidemiological studies. *Forensic Science Review, 27*(2), 89–114.

Gmel, G., Kuendig, H., Rehm, J., Schreyer, N., & Daeppen, J.-B. (2009). Alcohol and cannabis use as risk factors for injury--a case-crossover analysis in a Swiss hospital emergency department. BMC Public Health, 9(40), 1-9.

Gordis, L. (2014). *Epidemiology* (5th ed.). Philadelphia, PA: Elsevier/Saunders.

Gore, A. (2010) Know your limit: How Legislatures have gone overboard with per se drunk driving laws and how men pay the price. William & Mary Journal of Women and the Law, 16(2), 423-447.

Government of Canada. (2016). Access to Cannabis for Medical Purposes Regulations - Daily Amount Fact Sheet (Dosage). Retreived from https://www.canada.ca/en/health-canada/services/drugs-medication/cannabis/information-medical-practitioners/cannabis-medical-purposes-regulations-daily-amount-fact-sheet-dosage.html

Government of Canada. (2017). Backgrounder – Changes to Impaired Driving Laws. Retrieved from https://www.canada.ca/en/healthcanada/news/2017/04/backgrounder_changestoimpaireddrivinglaws.html

Grossman, M., Chaloupka, F. J., & Shim, K. (2002). Illegal drug use and public policy. *Health Affairs, 21*(2), 134-145.

Grotenhermen, F. (2003) Pharmacokinetics and pharmacodynamics of cannabinoids. *Clinical Pharmacokinetics, 42*(4), 328-360.

Grotenhermen, F., Leson, G., Berghaus, G., Drummer, O. H., Krüger, H.-P., & Longo, M., Moskowitz, H., Perrine, B., Ramaekers, J.G., Smiley, A. & Tunbridge, R. (2005). Developing Science-Based Per Se Limits for Driving under the Influence of Cannabis Findings and Recommendations by an Expert Panel. (DUIC Report) (September), 1–49.

Grotenhermen, F., Leson, G., Berghaus, G., Drummer, O.H., Krüger, H.P., Longo, M., Moskowitz, H., Perrine, B., Ramaekers, J.G., Smiley, A. & Tunbridge, R. (2007). Developing limits for driving under cannabis. *Addiction, 102*(12), 1910-1917.

Gruber, A. J., Pope, H. G., Hudson, J. I., & Yurgelun-Todd, D. (2003). Attributes of long-term heavy cannabis users: a case–control study. *Psychological Medicine, 33*(8), 1415-1422.

Gusfield, J.R. (1979). Managing competence: An ethnographic study of drinking-driving and the context of bars. In Hardford, T.C Harford, T. C., & Gaines, L. S (Eds). *Social drinking contexts: Proceedings of a workshop. Research monograph No7*. Superintendent of Documents, US Government Printing Office, Washington, DC.

Haeckel, R., & Peiffer, U. (1992). Comparison of ethanol concentration in saliva and blood from police controlled persons. *Blutalkohol, 29*(5), 342-349.

Haidich, A.B. (2010). Meta-analysis in medical research. *Hippokratia, 14*(1), 29–37.

Hall, W. (2014). What has research over the last two decades revealed about the adverse health effects of recreational cannabis use. *Addiction,* 110(1), 19–35.

Hallowell, Gerald (1988). Prohibition in Canada. In *The Canadian Encyclopedia*. Hurtig Publishers.

Hathaway, A. D, & Erickson, P. G. (2003). Drug Reform Principles and Policy Debates: Harm Reduction Prospects for Cannabis in Canada. *Journal of Drug Issue.* 33 (2): 465-496.

Harder, S., & Rietbrock, S. (1997) Concentration-effect relationship of Delta-9-tetrahydrocannabiol and prediction of psychotropic effects after smoking marijuana. *International Journal of Clinical Pharmacology and Therapeutics.* 35(4), 155-159

Hart, C.L., Van Gorp, W., Haney, M., Foltin, R.W., & Fischman, M.W. (2001). Effects of acute smoked marijuana on complex cognitive performance. *Neuropsychopharmacology, 25*(5), 757-765.

Hartman, R.L., & Huestis, M.A. (2013). Cannabis effects on driving skills. *Clinical Chemistry, 59*(3), 478–92.

Hartman, R. L., Anizan, S., Jang, M., Brown, T. L., Yun, K., Gorelick, D. A., & Huestis, M. A. (2015a). Cannabinoid disposition in oral fluid after controlled vaporizer administration with and without alcohol. *Forensic Toxicology, 33*(2), 260-278.

Hartman, R.L., Brown, T.L., Milavetz, G., Spurgin, A., Pierce, R.S., Gorelick, D., Gaffney, G., & Huestis, M.A. (2015b). Cannabis effects on driving lateral control with and without alcohol. *Drug and Alcohol Dependence, 154*, 25–37.

Hartman, R. L., Richman, J. E., Hayes, C. E., & Huestis, M. A. (2016). Drug Recognition Expert (DRE) examination characteristics of cannabis impairment. *Accident Analysis and Prevention*, 92, 219–229.

Hartman, R. L., Brown, T. L., Milavetz, G., Spurgin, A., Gorelick, D. A., Gaffney, G. R., & Huestis, M. A. (2016). Effect of blood collection time on measured δ9-Tetrahydrocannabinol concentrations: Implications for driving interpretation and drug policy. *Clinical Chemistry*, *62*(2), 367–377.

Hayden, J.W. (1991). Passive inhalation of marijuana smoke: a critical review. *Journal of Substance Abuse, 3*, 85-90.

Health Canada. (2012). *Canadian alcohol and drug use monitoring survey.* Retrieved from: https://www.canada.ca/en/health-canada/services/health-concerns/drug-prevention-treatment/drug-alcohol-use-statistics/canadian-alcohol-drug-use-monitoring-survey-summary-results-2012.html

Health Canada. (2013). *Information for health care professionals: cannabis (marihuana, marijuana) and the cannabinoids.* Retrieved from: http://www.hc-sc.gc.ca/dhp-mps/alt formats/pdf/marihuana/med/infoprof-eng.pdf

Health Canada. (2015). *Canadian tobacco alcohol and drugs survey (CTADS): 2015.* Retrieved from https://www.canada.ca/en/health-canada/services/canadian-tobacco-alcohol-drugs-survey/2015-summary.html

Health Canada. (undated). Health Effects of Cannabis. Retrieved from https://www.canada.ca/content/dam/hc-sc/documents/services/campaigns/27-16-1808-Factsheet-Health-Effects-eng-web.pdf

Heishman, S.J., Huestis, M.A., Henningfield, J., & Cone, E.J. (1990). Acute and residual effects of marijuana: profiles of plasma thc levels, physiological, subjective and performance measures. *Pharmacology Biochemistry & Behavior, 37*(3), 561–565.

Hels, T., Lyckegaard, A., Simonsen, K.W., Steentoft, A., & Bernhoft, I.M. (2013). Risk of severe driver injury by driving with psychoactive substances. *Accident; Analysis and Prevention, 59*, 346–56.

Hostiuc, S., Moldoveanu, A., Negoi, I., & Drima, E. (2018). The Association of Unfavorable Traffic Events and Cannabis Usage: A Meta-Analysis. *Frontiers in Pharmacology*, 9(February).

Howland, J., Rohsenow, D.J., & Greece J.A. (2010). The effects of binge drinking on college students' next-day academic test-taking performance and mood state. *Addiction*, 105: 655-65.

Huestis, M. A., Henningfield, J. E., & Cone, E. J. (1992). Blood Cannabinoids. I. Absorption of thc and formation of 11-OH-THC and THCCOOH during and after smoking marijuana. *Journal of Analytical Toxicology*, 16(5), 276–282.

Huestis, M. A., & Cone, E. (1995). Drug test findings resulting from unconventional drug exposure. In *Handbook of Workplace Drug Testing* (pp.289-320). Washington, DC: AACC Press

Huestis, M.A, Mitchell, J.M., & Cone, EJ. (1995). Detection times of marijuana metabolites in urine by immunoassay and GC-MS. *Journal of Analytical Toxicology, 19*(6), 443–9.

Huestis, M.A, & Cone, E.J. (2004). Relationship of Delta 9-tetrahydrocannabinol concentrations in oral fluid and plasma after controlled administration of smoked cannabis. *Journal of Analytical Toxicology, 28*(6), 394–9.

Huestis, M.A., & Cone, E.J. (2007). Methamphetamine disposition in oral fluid, plasma, and urine. *Annals of the New York Academy of Sciences, 1098*, 104–121.

Hunault, C.C., Böcker, K.B.E., Stellato, R.K., Kenemans, J.L., de Vries, I., & Meulenbelt, J. (2014). Acute subjective effects after smoking joints containing up to 69 mg Δ9-tetrahydrocannabinol in recreational users: a randomized, crossover clinical trial. *Psychopharmacology, 231*(24), 4733.

Insurance Institute for Highway Safety (IIHS). (2013). *Fatality facts: Teenagers 2013.* Retrieved from: http://www.iihs.org/iihs/topics/t/teenagers/fatalityfacts/teenagers\

International Alliance for Responsible Drinking. (2017). *Drinking Guidelines: General Population.* Retrieved from http://www.iard.org/policy-tables/drinking-guidelines-general-population/

International Center for Alcohol Policies. (1998). *What is a "Standard Drink" (Report No. 5).* Retrieved from https://web.archive.org/web/20160305015945/http://icap.org/portals/0/download/all pdfs/icap reports english/report5.pdf

International Transport Forum. (2016). *Road Safety Annual Report 2016.* Retrieved from https://www.itf-oecd.org/road-safety-annual-report-2016

Iversen, L. (2003). Cannabis and the brain. *Brain, 126*(6), 1252–1270.

Jaffee, W. B., Trucco, E., Levy, S., & Weiss, R. D. (2007). Is this urine really negative? A systematic review of tampering methods in urine drug screening and testing. *Journal of Substance Abuse Treatment, 33*(1), 33–42.

Jaffee, W.B., Trucco, E., Teter, C., Levy, S., & Weiss, R.D. (2008). Ensuring Validity in Urine Drug Testing. *Psychiatric Services, 59,* 140-142.

Jaffe, D. H., Siman-Tov, M., Gopher, A., & Peleg, K. (2013). Variability in the blood/breath alcohol ratio and implications for evidentiary purposes. *Journal of Forensic Sciences, 58*(5), 1233–1237.

Jellinek, E.M., & McFarland, R.A. (1940). Analysis of psychological experiments on the effects of alcohol. *Quarterly Journal of Alcohol Studies, 1,* 272-371.

Jiang, Y. (1994). Punishment celerity and severity: Testing a specific deterrence model on drunk driving recidivism. *Journal of Criminal Justice, 22,* 355-66.

Jonah, B., Yuen, L., Au-Yeung, E., Paterson, D., Dawson, N., Thiessen, R., & Arora, H. (1999). Front-line police officers' practices, perceptions and attitudes about the enforcement of impaired driving laws in Canada. *Accident Analysis and Prevention, 31* (5), 421–443.

Jones, A. W. (1995). Measuring ethanol in saliva with the QED® enzymatic test device: comparison of results with blood-and breath-alcohol concentrations. *Journal of analytical toxicology, 19*(3), 169-174.

Jones, A.W., & Andersson, L. (1996). Variability of the blood/breath alcohol ratio in drinking drivers. *Journal of Forensic Science 41*(6): 916-921.

Jones, A. W. (2010). *Road Safety Web Publication No. 15. The Relationship between Blood Alcohol Concentration (BAC) and Breath Alcohol Concentration (BrAC): A Review of the Evidence.* Department for Transport. London. Retrieved from www.dft.gov.uk/pgr/road-safety/research/rsrr

Jones, B. M., & Vega, A. (1972). Cognitive performance measured on the ascending and descending limb of the blood alcohol curve. *Psychopharmacologia, 23*(2), 99–114.

Kadehjian, L. (2005). Legal issues in oral fluid testing. *Forensic Science International, 150,* 151-160.

Kapur, B. (1994). Drug testing methods and interpretations of drug testing results. In S. Macdonald and P. Roman (Eds.), *Research Advances in Alcohol and Drug Problems.* New York: Plenum Press.

Kapur, B. (2009) Drug testing methods and clinical interpretation of test results. *Encyclopedia of Drugs, Alcohol and Addictive Behavior* (3rd ed.), (Eds. H. Kranzler and P. Korsmeyer) Volume 1. Farmington Hills: Gale Cengage Learning.

Karschner, E., Schwilke, E., & Lowe, R. (2009). Do δ9-tetrahydrocannabinol concentrations indicate recent use in chronic cannabis users? *Addiction, 104*(12), 2041–2048.

Karschner, E. L., Swortwood, M. J., Hirvonen, J., Goodwin, R. S., Bosker, W. M., Ramaekers, J. G., & Huestis, M. A. (2015). Extended plasma cannabinoid excretion in chronic frequent cannabis smokers during sustained abstinence and correlation with psychomotor performance. *Drug Testing and Analysis 8*(7), 682-689.

Kay, G. G., & Logan, B. K. (2011). *Drugged driving expert panel report: a consensus protocol for assessing the potential of drugs to impair driving. (DOT HS 811 438).* Washington: DC, National Highway Traffic Safety Administration.

Keeping, Z., & Huggins, R. (undated). *Final Report on the Oral Fluid Drug Screening Device Pilot Project. Public Safety Canada Royal Canadian Mounted Police Canadian Council of Motor Transport Administrators.* Retrieved from https://www.publicsafety.gc.ca/cnt/rsrcs/pblctns/rl-fld-drg-scrnng-dvc-plt/index-en.aspx

Kelley-Baker, T., Moore, C., Lacey, J.H., & Yao, J. (2014). Comparing drug detection in oral fluid and blood: data from a national sample of nighttime drivers. *Traffic Injury Prevention, 15*(2), 111–8.

Kelly, D. (2014) 10 Outrageous claims made by the Temperance Movement, *Listverse,* Retrieved from https://listverse.com/2014/01/05/10-outrageous-claims-made-by-the-temperance-movement/

Kelly, E., Darke, S., & Ross, J. (2004). A review of drug use and driving: epidemiology, impairment, risk factors and risk perceptions. *Drug and Alcohol Review, 23*(3), 319–44.

Kelsey, J. L., Whittemore, A. S., Alfred, E. S., & Thompson, D. W. (1996). Case Control Studies Planning and Execution. In *Methods in Observational Epidemiology,* (2nd ed., pp. 188–213). New York: Oxford University Press.

Kenkel, J.F. (2015). *Impaired Driving in Canada* (4th ed). Markham, Ont: LexisNexis Canada.

Khiabani, H.Z., Bramness, J.G., Bjørneboe, A., & Mørland, J. (2006). Relationship between THC concentration in blood and impairment in apprehended drivers. *Traffic Injury Prevention, 7*(2), 111–6.

Kim, D-J., Yoon, S.-J., Lee, H.P., Choi, B.-M., & Go, H. J. (2003). The effects of alcohol hangover on cognitive functions in healthy subjects. *International Journal of Neuroscience, 113*(4), 581–594.

Kim, I., Barnes, A.J., Oyler, J.M., Schepers, R., Joseph, R.E., Cone, E.J., Lafko, D., Moolchan, E. T., & Huestis, M.A. (2002). Plasma and oral fluid pharmacokinetics and pharmacodynamics after oral codeine administration. *Clinical Chemistry, 48*(9), 1486–1496.

Klette, H. (1983) Drinking and driving in Sweden—the law and its effects Lund University, Lund, Sweden.

Krotulski, A. J., Mohr, A. L. A., Friscia, M., & Logan, B. K. (2017). Field detection of drugs of abuse in oral fluid using the Alere™ DDS®2 mobile test system with confirmation by liquid chromatography tandem mass spectrometry (LC–MS/MS). *Journal of Analytical Toxicology*, 1–7.

Kuypers, K.P.C., Legrand, S.A., Ramaekers, J.G., & Verstraete, A.G. (2012). A case-control study estimating accident risk for alcohol, medicines and illegal drugs. *PloS One*, *7*(8), e43496–e43496.

Lacey, J.H., Kelley-Baker, T., Voas, R.B., Romano, E., Furr-Holden, C.D., Torres, P., & Berning, A. (2011). Alcohol and drug-involved driving in the United States: methodology for the 2007 National Roadside Survey. *Evaluation Review, 35*(4), 319–53.

Lacey, J. H., Kelley-Baker, T., Berning, A., Romano, E., Ramirez, A., Yao, J., ...& Compton, R. (2016, December). Drug and alcohol crash risk: A case-control study (Report No. DOT HS 812 355). Washington, DC: National Highway Traffic Safety Administration.

Langel, K., Gjerde, H., Favretto, D., Lillsunde, P., Øiestad, E.L., Ferrara, S.D., & Verstraete, A.G. (2014). Comparison of drug concentrations between whole blood and oral fluid. *Drug Testing and Analysis*, 461–471.

Last, J. M. (1995). *A dictionary of epidemiology* (3rd ed.). New York: Oxford University Press.

Laumon, B., Gadegbeku, B., Martin, J.L., & Biecheler, M.B. (2005). Cannabis intoxication and fatal road crashes in France: population based case-control study. *British Medical Journal (Clinical Research Ed.) 331*(7529), 1371.

Lee, D., Schwope, D. M., Milman, G., Barnes, A. J., Gorelick, D. a, & Huestis, M. a. (2012). Cannabinoid disposition in oral fluid after controlled smoked cannabis. *Clinical Chemistry*, *58*(4), 748–756.

Leirer, V.O., Yesavage, J.A., & Morrow, D.G. (1989). Marijuana, aging, and task difficulty effects on pilot performance. *Aviation, Space, and Environmental Medicine*, 60(12), 1145–52.

Leirer, V., Yesavage, J., & Morrow, D. (1991). Marijuana carry-over effects on aircraft pilot performance. *Aviation, Space and Environmental Medicine*, 62(3), 221–227.

Lenné, M.G., Dietze, P.M., Triggs, T.J., Walmsley, S., Murphy, B., & Redman, J.R. (2010). The effects of cannabis and alcohol on simulated arterial driving: Influences of driving experience and task demand. *Accident Analysis & Prevention*, 42(3), 859–66.

Lerner, B. H. (2012). Drunk driving across the globe: Let's learn from one another. *The Lancet, 379*(9829), 1870-1871.

Levin, K. H., Copersino, M. L., Heishman, S. J., Liu, F., Kelly, D. L., Boggs, D. L., & Gorelick, D. A. (2010). Cannabis withdrawal symptoms in non-treatment-seeking adult cannabis smokers. *Drug and Alcohol Dependence*, 111(1–2), 120–7.

Li, G., Brady, J. E., DiMaggio, C., Baker, S. P., & Rebok, G. W. (2010). Validity of suspected alcohol and drug violations in aviation employees. *Addiction, 105*(10), 1771–5.

Li, G., Baker, S. P., Zhao, Q., Brady, J. E., Lang, B. H., Rebok, G. W., & DiMaggio, C. (2011). Drug violations and aviation accidents: findings from the US mandatory drug testing programs. *Addiction, 106*(7), 1287–92.

Li, G., Brady, J. E., & Chen, Q. (2013). Drug use and fatal motor vehicle crashes: a case-control study. *Accident Analysis and Prevention*, 60, 205–10.

Li, M.-C., Brady, J. E., DiMaggio, C. J., Lusardi, A. R., Tzong, K. Y., & Li, G. (2012). Marijuana use and motor vehicle crashes. *Epidemiologic Reviews*, 34(1), 65–72.

Ling, J., Stephens, R., & M Heffernan, T. (2010). Cognitive and psychomotor performance during alcohol hangover. *Current drug abuse reviews, 3*(2), 80-87.

Lindberg, L., Brauer, S., Wollmer, P., Goldberg, L., Jones, A.W., & Olsson, S.G. (2007). Breath alcohol concentration determined with a new analyzer using free exhalation predicts almost precisely the arterial blood alcohol concentration. *Forensic Science International*, 168(2-3), 200–7.

Longo, M.C., Hunter, C.E., Lokan, R.J., White, J.M., & White, M.A. (2000b). The prevalence of alcohol, cannabinoids, benzodiazepines and stimulants amongst injured drivers and their role in driver culpability. Part II: The relationship between drug prevalence and drug concentration, and driver capability. *Accident Analysis and Prevention*, 32, 623-632.

Lopez-Quintero, C., Cobos, J.P. de los, Hasin, D.S., Okuda, M., Wang, S., Grant, B.F., & Blanco, C. (2011). Probability and predictors of transition from first use to dependence on nicotine, alcohol, cannabis, and cocaine: Results of the National Epidemiologic Survey on Alcohol and Related Conditions (NESARC). *Drug and Alcohol Dependence*, 115(1-2), 120–130.

Macdonald, S., & Pederson, L. L. (1988). Occurrence and patterns of driving behaviour for alcoholics in treatment. *Drug and Alcohol Dependence*, 22(1–2).

Macdonald, S. (1989). A comparison of the psychosocial characteristics of alcoholics responsible for impaired and nonimpaired collisions. Accident Analysis and Prevention, 21(5), 493-508.

Macdonald, S., & Wells, S. (1994). The impact and effectiveness of drug testing programs in the workplace. In S. Macdonald & P. Roman (Eds.), *Drug testing in the workplace* (pp. 121–140). New York: Plenum Press.

Macdonald, S. & Roman, P. (Eds.). (1994). *Drug testing in the workplace.* New York: Plenum Press.

Macdonald, S. (1995). The role of drugs in workplace injuries: Is drug testing appropriate? *Journal of Drug Issues*, 25(4), 703-722.

Macdonald, S. (2003). The influence of the age and sex distributions of drivers on the reduction of impaired crashes: Ontario, 1974-1999. *Traffic Injury Prevention*, 4(1), 33–37.

Macdonald, S., Mann, R., Chipman, M., Pakula, B., Erickson, P., Hathaway, A., & MacIntyre, P. (2008). Driving behavior under the influence of cannabis or cocaine. *Traffic Injury Prevention*, 9(3), 190–194.

Macdonald, S., Hall, W., Roman, P., Stockwell, T., Coghlan, M., & Nesvaag, S. (2009). Testing for cannabis in the workplace: A review of the evidence. *Addiction, 105*(3), 408-416.

Macdonald, S., Zhao, J., Martin, G., Brubacher, J., Stockwell, T., Arason, N., Chan, H. (2013). The impact on alcohol-related collisions of the partial decriminalization of impaired driving in British Columbia, Canada. *Accident; Analysis and Prevention*, *59*, 200–205.

Macdonald, S., Stockwell, T., Reist, D., Belle-Isle, L., Benoit, C., Callaghan, R., Cherpitel, C., Dyck, T., Jansson, M., Pauly, B., Roth, E., Vallance, K. & Zhao, J. (2016). Legalization of Cannabis in Canada: Implementation strategies and public health. CARBC Bulletin #16, Victoria, BC: University of Victoria.

Macdonald, S. (2018). *Essentials of statistics with SPSS - with examples from the Canadian Community Health Survey* (2nd ed). Lulu.

Macdonald, S., Greer, A. & Ferencz, S. Police carding and discretion with drug using youth in British Columbia (BC), Canada. The International Society for the Study of Drug Policy, Vancouver, B.C., May 15-18, 2018.

Mann, R. E., Macdonald, S., Stoduto, L. G., Bondy, S., Jonah, B., & Shaikh, A. (2001). The effects of introducing or lowering legal per se blood alcohol limits for driving: an international review. *Accident Analysis & Prevention, 33*(5), 569–83.

Mann, R. (2002). Choosing a rational threshold for the definition of drunk driving: what research recommends. *Addiction*, 97, 1237–1238.

Mann, R. E., Brands, B., Macdonald, S., & Stoduto, G. (2003). *Impacts of cannabis on driving: An analysis of current evidence with an emphasis on Canadian data*. Transport Canada, Road Safety and Motor Vehicle Regulation Directorate.

Mann, R. E., Stoduto, G., Ialomiteanu, A., Asbridge, M., Smart, R. G., & Wickens, C. M. (2010). Self-reported collision risk associated with cannabis use and driving after cannabis use among Ontario adults. *Traffic Injury Prevention*, *11*(2), 115–22.

Marquet, P., Delpla, P., Kerguelen, S., Bremond, P., & Al. E. (1998). Prevalence of drugs of abuse in urine of drivers involved in road accidents in France: A collaborative study. *Journal of Forensic Science*, *43*(4), 806–811.

Martin, C.S. (1998) Alcohol. in Karsh, S. (Eds) Drug Abuse Handbook. Washington, CRC Press.

Martin, T. L., Solbeck, P. A., Mayers, D. J., Langille, R. M., Buczek, Y., & Pelletier, M. R. (2013). A review of alcohol-impaired driving: The role of blood alcohol concentration and complexity of the driving task. *Journal of Forensic Sciences*, *58*(5), 1238-1250.

McKim, W. (1986). *Drugs and behavior.* Prentice-Hall, New Jersey.

McKinney, A., Kieran, C., & Verster, J. (2012). Direct comparison of the cognitive effects of acute alcohol with the morning after a normal night's drinking. Human *Psychopharmacology*, 27(3), 295–304.

McLaren, J., Swift, W., Dillon, P., & Allsop, S. (2008). Cannabis potency and contamination: a review of the literature. *Addiction, 103*(7), 1100–1109.

McLellan, A. A., Ware, M. A., Boyd, S., Chow, G., Jesso, M., Kendall, P... & Zahn, C. (2016). *A framework for the legalization and regulation of cannabis in Canada: The final report of the task force on cannabis legalization and regulation.* Ottawa: Health Canada.

Ménétrey, A., Augsburger, M., Favrat, B., Pin, M. a, Rothuizen, L. E., Appenzeller, M., Giroud, C. (2005). Assessment of driving capability through the use of clinical and psychomotor tests in relation to blood cannabinoids levels following oral administration of 20 mg dronabinol or of a cannabis decoction made with 20 or 60 mg Delta9-THC. *Journal of Analytical Toxicology*, *29*(5), 327–338.

Miller, T., Blewden, M., & Zhang, J. (2004). Cost savings from a sustained compulsory breath testing and media campaign in New Zealand. *Accident Analysis and Prevention*, *36*(5), 783–794.

Milman, G., Schwope, D.M., Gorelick, D.A, & Huestis, M.A. (2012). Cannabinoids and metabolites in expectorated oral fluid following controlled smoked cannabis. *Clinica Chimica Acta, 413*(7-8), 765–70.

Moeller, K.E., Lee, K.C., & Kissack, J.C. (2008). Urine drug screening: practical guide for clinicians. *Mayo Clinic Proceedings*, 83(1), 66-76.

Moore, C., Coulter, C., Uges, D., Tuyay, J., van der Linde, S., van Leeuwen, A., Garnier, M., & Orbita, J. (2011). Cannabinoids in oral fluid following passive exposure to marijuana smoke. *Forensic Science International*, 212(1-3), 227–230.

Moskowitz, H. (1973). Laboratory studies of the effects of alcohol on some variables related to driving. *Journal of Safety Research, 5*(3), 185-199.

Moskowitz, H. (1985). Marihuana and Driving. *Accident Analysis & Prevention, 17*(4), 323–345.

Moskowitz, M, Florentino, D. (2000). *A review of the literature on the effects of low doses of alcohol on driving-related skills*. Washington, DC.

Movig, K. L. L., Mathijssen, M. P. M., Nagel, P. H. A, van Egmond, T., de Gier, J. J., Leufkens, H. G. M., & Egberts, a C. G. (2004). Psychoactive substance use and the risk of motor vehicle accidents. *Accident Analysis & Prevention, 36*(4), 631–6.

Mule, S.J., & Casella, G.A. (1988). Active and realistic passive marijuana exposure tested by three immunoassays and GC/MS in urine. *Journal of Analytical Toxicology*, 12, 113-116.

Mura, P., Kintz, P., Dumestre, V., Raul, S., & Hauet, T. (2005). THC can be detected in brain while absent in blood. *Journal of Analytical Toxicology*, 29(8), 842–3.

Mura, P., Kintz, P., Ludes, B., Gaulier, J. M., Marquet, P., Martin-Dupont, S., Vincent, F., Kaddour, A., Goullé, J.P., Nouveau, J., Mouslma, M., Tilhet-Coartet, S., & Pourrat, O. (2003). Comparison of the prevalence of alcohol, cannabis and other drugs between 900 injured drivers and 900 control subjects: results of a French collaborative study. *Forensic Science International, 133*(1-2), 79–85.

Nagin D.S., & Pogarsky G. (2001). Integrating celerity, impulsivity, and extralegal sanction threats into a model of general deterrence: Theory and evidence. *Criminology (2001)* 39, 865-92.

National Center for Statistics and Analysis. (2017, October). Alcohol impaired driving: 2016 data (Traffic Safety Facts. Report No. DOT HS 812 450). Washington, DC: National Highway Traffic Safety Administration.

National Institute of Justice (NIJ). (2014). *Five Things About Deterrence.* Retrieved from https://nij.gov/five-things/pages/deterrence.aspx

Neavyn, M. J., Blohm, E., Babu, K. M., & Bird, S. B. (2014). Medical marijuana and driving: A review. *Journal of Medical Toxicology*, 269–279.

Newmeyer, M.N., Desrosiers, N.A, Lee, D., Mendu, D.R., Barnes, A.J., Gorelick, D.A, & Huestis, M.A. (2014). Cannabinoid disposition in oral fluid after controlled cannabis smoking in frequent and occasional smokers. *Drug Testing and Analysis*, 6(10), 1002–10.

Niedbala, R. S., Kardos, K.W., Fritch, D.F., Kardos, S., Fries, T., Waga, J., Robb, J., & Cone, E.J. (2001) Detection of marijuana use by oral fluid and urine analysis following single-dose administration of smoked and oral marijuana. *Journal of Analytical Toxicology*, 25, 289-303.

Niedbala, S. Karados, K.W., Fritch, D.F., Kunsman, K.P., Blum, K.A., Newland, G.A., Waga, J., Kurtz, L., Bronsgeest, M. & Cone, E.J. (2005). Passive cannabis smoke exposure and oral fluid testing II: two studies of extreme cannabis smoke exposure in a motor vehicle. *Journal of Analytical Toxicology*, *29*, 607-615.

Niedbala, S., Karados, K., Salamone, S., Fritch, D., Bronsgeest, M., & Cone, E.J. (2004). Passive cannabis exposure and oral fluid testing. *Journal of Analytical Toxicology*, *28*, 546-552.

Nolin, P. C. & Kenny, C. (2002). *Cannabis: our position for a Canadian public policy: report of the Senate Special Committee on Illegal Drugs*. Senate Canada.

Nordqvist, C. (2018). What is alcohol abuse disorder, and what is the treatment? Medical News Today. May 29. Retrieved from https://www.medicalnewstoday.com/articles/157163.php

Nutt, D., King, L. & Phillips, L. (2010). Drug harms in the UK: a multicriteria decision analysis. *Lancet*, 376, 1558–1565.

Ogden, E. J. D., & Moskowitz, H. (2004). Effects of alcohol and other drugs on driver performance. *Traffic Injury Prevention*, 5(3), 185–98.

Okrent, D. (2010). *Last call: The rise and fall of prohibition.* Simon and Schuster.

O'Kane, C. J., Tutt, D. C., & Bauer, L. A. (2002). Cannabis and driving: a new perspective. *Emergency Medicine*, 14(3), 296–303.

Ontario Law Reform Commission. (1992). *Report on drug and alcohol testing in the workplace.* Toronto: Law Reform Commission.

Ossola, A. (2015). *A breathalyzer test for all sorts of drugs.* Popular Science. Retrieved from http://www.popsci.com/breathalyzer-drugs

Owusu-Bempah, A. (2014). Cannabis impaired driving: An evaluation of current modes of detection. *Canadian Journal of Criminology and Criminal Justice*, *56*(2), 219-240.

Papafotiou, K., Carter, J. D., & Stough, C. (2005). An evaluation of the sensitivity of the Standardised Field Sobriety Tests (SFSTs) to detect impairment due to marijuana intoxication. *Psychopharmacology*, 180(1), 107–114.

Paternoster, R. (2010). How Much Do We Really Know about Criminal Deterrence. *Journal of Criminal Law and Criminology*, 100(3), 765–823.

Payne, J. P., Foster, D. V, Hill, D. W., & Wood, D. G. (1967). Observations on interpretation of blood alcohol levels derived from analysis of urine. *British Medical Journal*, 3(5569), 819–823.

Peck, R. C., Gerbers, R. B., Voas, E., & Romano, E. (2007). Improved methods for estimating relative crash risk in a case-control study of blood alcohol levels. In *International Conference on Alcohol Drugs and Traffic Safety.*

Perreault, S. (2016). *Impaired Driving in Canada, 2015.* Canadian Centre for Justice Statistics, Statistics Canada.

Pil, K., & Verstraete, A. (2008). Current developments in drug testing in oral fluid. *Therapeutic Drug Monitoring*, *30*(2), 196–202.

Platt, B. (2018). How much cannabis could you smoke and stay under the proposed legal limit for driving? The answer may be zero. *National Post.* June 5. Retrieved from https://nationalpost.com/news/politics/how-much-cannabis-is-safe-to-consume-before-driving-nobody-knows-but-were-setting-legal-limits-anyway

Polinsky AM, Shavell S. The Economic Theory of Public Enforcement of Law. *Journal of Economic Literature*, 2000(38), 45-76.

Pope, H., Gruber, A., Hudson, J., Huestis, M., & Yurgelun-Todd, D. (2002). Cognitive measures in long-term cannabis users. *The Journal of Clinical Pharmacology*, *42*(11), 41–47.

Porath-Waller, A.J., Bierness, D.J., & Beasley, E.E. (2009). Toward a more parsimonious approach to drug recognition expert evaluations. *Traffic Injury Prevention*, 10, 513-518.

Prashad, S., & Filbey, F. M. (2017). Cognitive motor deficits in cannabis users. *Current Opinion in Behavioral Sciences*, 13, 1–7.

Rainey, P. M. (1993). Relation between serum and whole blood ethanol concentrations. *Clinical Chemistry*, 39(11), 2288–2292.

Ramaekers, J., Berghaus, G., van Laar, M., & Drummer, O. (2004). Dose related risk of motor vehicle crashes after cannabis use. *Drug and Alcohol Dependence*, 73(2), 109–119.

Ramaekers, J.G., Berghaus, G., van Laar, M., & Drummer, O.H. (2009) Dose related risk of motor vehicle crashes after cannabis use: an update. In: Verster J.C., Pandi-Perumal S.R., Ramaekers J.G., and de Gier J.J. (Eds.), *Drugs, Driving and Traffic Safety*. Birkhäuser Basel.

Ramaekers, J.G., Kauert, G., Theunissen, E.L., Toennes, S.W., & Moeller, M.R. (2009). Neurocognitive performance during acute THC intoxication in heavy and occasional cannabis users. *Journal of Psychopharmacology*, 23(3), 266–77.

Ramaekers, J.G., Moeller, M.R., van Ruitenbeek, P., Theunissen, E.L., Schneider, E., & Kauert, G. (2006). Cognition and motor control as a function of Delta9-THC concentration in serum and oral fluid: limits of impairment. *Drug and Alcohol Dependence*, 85(2), 114–22.

Reilly, T., & Scott, J. (1993). Effects of elevating blood alcohol levels on tasks related to dart throwing. *Perceptual and Motor Skills*, *77*(1), 25–26.

Remington, P.L., Brownson, R.C. & Wegner, M.V. (2010). *Chronic Disease Epidemiology and Control* (3rd Ed.). Washington DC, APHA Press.

Road Safety and Motor Vehicle Regulation Directorate. (2011). *Alcohol use by drivers fatally injured in motor vehicle collisions in Canada in 2008 and the previous 21 years.* Ottawa.

Robbe, H.W.J, & O'Hanlon, J.F. (1993). *Marijuana and actual driving performance: Final report.* US Department of Transportation, National Highway Traffic Safety Administration.

Rogeberg, O., & Elvik, R. (2016). The effects of cannabis intoxication on motor vehicle collision revisited and revised. *Addiction*, *111*(8), 1348-1359.

Ronen, A., Gershon, P., Drobiner, H., Rabinovich, A., Bar-Hamburger, R., Mechoulam, R., Cassuto, Y., & Shinar, D. (2008). Effects of THC on driving performance, physiological state and subjective feelings relative to alcohol. *Accident Analysis & Prevention*, 40(3), 926–34.

Rosner, B. (1982) *Fundamentals of Biostatistics.* Boston, MA: Duxbury Press.

Ross, H. L., Campbell, D. T., & Glass, G. V. (1970). Determining the social effects of a legal reform: The British "breathalyser" crackdown of 1967. *American Behavioral Scientist*, *13*(4), 493-509.

Ross, H. (1975) The Scandinavian Myth: The Effectiveness of Drinking-and-Driving Legislation in Sweden and Norway. The Journal of Legal Studies, 4(2), 285-310.

Ross, H. L., Campbell, D.T. & Glass, G.V. (1975). The Effectiveness of Drinking-and Driving Laws in Sweden and Great Britain. *Toxicomanies*, 663–678.

Ross, H.L. (1982*). Deterring the Drinking Driver - Legal Policy and Social Control.* New York: Lexington Books.

Ross, H. L. (1984). Social control through deterrence: Drinking and Driving Laws. *Annual Review of Sociology*, 21–35.

Ross, H. L. & Klette, H. (1995). Abandonment of mandatory jail for impaired drivers in Norway and Sweden. *Accident Analysis & Prevention,* 27(2), 151-157.

Ross, H. L. (2017). Social control through deterrence : drinking-and-driving laws. *Annual Review of Sociology*, *10* (1984), 21–35.

Royal Canadian Mounted Police, (undated). Drug Recognition Expert Evaluations, Retrieved from http://www.rcmp-grc.gc.ca/ts-sr/dree-eert-eng.htm.

Saila, J. (2009) *Comparing the concentrations of drugs and medicines in whole blood, plasma and oral fluid samples of drivers suspected of driving under the influence*. (Thesis in Biomedical Laboratory Science, Bachelor of Health Care). Helsinki Metropolia University of Applied Sciences.

Sauber-Schatz, E.K., Ederer, D.J., Dellinger, A.M., & Baldwin, G.T. (2016). Vital signs: motor vehicle injury prevention—United States and 19 comparison countries. *Morbidity and Mortality Weekly Report,* 65.

Scheidweiler, K. B., Spargo, E. A. K., Kelly, T. L., Cone, E. J., Barnes, A. J., & Huestis, M. A. (2010). Pharmacokinetics of Cocaine and Metabolites in Human Oral Fluid and Correlation with Plasma Concentrations following Controlled Administration. *Therapeutic Drug Monitoring*, 32(5), 628–37.

Shadish, W.R, Cook, T.D & Campbell, D.T. (2002). *Experimental and quasi-experimental designs for generalized causal inference.* Boston, MA: Houghton Mifflin.

Sharma, P., Murthy, P., & Bharath, M. M. S. (2012). Chemistry, Metabolism, and Toxicology of Cannabis: Clinical Implications. *Iran Journal of Psychiatry, 7*(14), 149–156.

Schwilke, E. W., Karschner, E. L., Lowe, R. H., Gordon, A. M., Lud, J., Herning, R. I., & Huestis, M. A. (2011). Intra- and intersubject whole blood/plasma cannabinoid ratios determined by 2-dimensional, electron impact gc-ms with cryofocusing. *Clinical Chemistry, 55*(6), 1188–1195.

Schwope, D. M., Bosker, W. M., Ramaekers, J. G., Gorelick, D. A., & Huestis, M. A. (2012). Psychomotor Performance, Subjective and Physiological Effects and Whole Blood Delta 9 -Tetrahydrocannabinol Concentrations in Heavy, Chronic Cannabis Smokers Following Acute Smoked Cannabis. *Journal of Analytical Toxicology*, *36*(May), 405–412.

Searle, J. (2015). Alcohol calculations and their uncertainty. *Medicine, Science and the Law, 55*(1), 58–64.

Semeniuk, I. (2018) Survey highlights tensions in public attitudes towards science and technology. *The Globe and Mail.* September 17 (A4).

Shults, R. a, Elder, R. W., Sleet, D. a, Nichols, J. L., Alao, M. O., Carande-Kulis, V. G., ... Thompson, R. S. (2001). Reviews of evidence regarding interventions to reduce alcohol-impaired driving. American Journal of Preventive Medicine, 21(4), 66–88.

Simpson, H. M., Beirness, D. J., Robertson, R. D., Mayhew, D. R., & Hedlund, J. H. (2004). Hard core drinking drivers. *Traffic injury prevention, 5*(3), 261-269.

Simonsen, K.W., Steentoft, A., Hels, T., Bernhoft, I.M., Rasmussen, B.S., & Linnet, K. (2012). Presence of psychoactive substances in oral fluid from randomly selected drivers in Denmark. *Forensic Science International, 221*(1-3), 33–8.

Smiley, A. (1999). Marijuana: On-road and driving-simulator studies. In: H. Kalant, W. Corrigall, W. Hall, & R. Smart. (Eds.), *The health effects of cannabis* (pp. 173-191). Toronto, ON: Addiction Research Foundation.

Smith, J.A., Hayes, C.E., Yolton, R.L., Rutledge, D.A., & Citek, K. (2002). Drug recognition expert evaluations made using limited data. *Forensic Science International, 130*, 167-173.

Smith, N. T. (2002). A review of the published literature into cannabis withdrawal symptoms in human users. *Addiction*, 97(6), 621–32.

Smith-Kielland, A., Skuterud, B., & Mørland, J. (1999). Urinary excretion of 11-nor-9-carboxy-delta9-tetrahydrocannabinol and cannabinoids in frequent and infrequent drug users. Journal of Analytical Toxicology, 23(5), 323–332. https://doi.org/10.1093/jat/23.5.323

Snowden, C. B., Miller, T. R., Waehrer, G. M., & Spicer, R. S. (2007). Random alcohol testing reduced alcohol-involved fatal crashes of drivers of large trucks. *Journal of Studies on Alcohol and Drugs*, 68(5), 634–40.

Solomon, R., & Chamberlain, E. (undated). An overview of Federal Drug-impaired Driving Enforcement and Provincial Licence Suspensions in Canada. MADD Canada.

Solomon, R., Chamberlain, E., Abdoullaeva, M., & Tinholt, B. (2011). Random breath testing: a Canadian perspective. *Traffic Injury Prevention*, 12(2), 111–119.

Solowij, N., Michie, P.T., & Fox, A.M. (1995). Differential impairments of selective attention due to frequency and duration of cannabis use. *Biological Psychiatry, 37*(10), 731–739.

Stafford M.C., & Warr, M. A. (1993). Reconceptualization of general and specific deterrence. *Journal of research in crime and delinquency,* 123-35.

Statistics Canada (2018). *National Cannabis Survey, Second Quarter 2018.* https://www150.statcan.gc.ca/n1/daily-quotidien/180809/dq180809a-eng.htm

Stavro, K., Pelletier, J., & Potvin, S. (2013). Widespread and sustained cognitive deficits in alcoholism: a meta-analysis. *Addiction Biology, 18*(2), 203–13.

Stephanson, N., Sandqvist, S., Lambert, M. S., & Beck, O. (2015). Method validation and application of a liquid chromatography-tandem mass spectrometry method for drugs of abuse testing in exhaled breath. *Journal of Chromatography* B: Analytical Technologies in the Biomedical and Life Sciences, 985, 189–196.

Stephens, R., Ling, J., Heffernan, T. M., Heather, N., & Jones, K. (2008). A review of the literature on the cognitive effects of alcohol hangover. *Alcohol and Alcoholism*, 43(2), 163–170.

Stuster, J. (2006). Validation of the standardized field sobriety test battery at 0.08% blood alcohol concentration. *Human factors, 48*(3), 608-614.

Stuster, J. W., & Burns, M. (1998). Validation of the Standardized Field Sobritiey Tests Battery at BSCs below .1% alcohl, (DOT HS 808 839), National Highway Traffic Safety Administration, Washington, D.C.

Swedler, D.I., Bowman, S.M. and Baker, S.P. (2012). Gender and age differences among teen drivers in fatal crashes. *Annals of the Association for the Advancement of Automotive Medicine*, 56, 97-106.

Swontinsky, R.B. (2015). *The Medical Review Officers Manual: MROCC'S Guide to Drug Testing*. Beverly Frams, MA: OEM Press.

Taylor, B., Irving, H. M., Kanteres, F., Room, R., Borges, G., Cherpitel, C., ... Rehm, J. (2010). The more you drink, the harder you fall: a systematic review and meta-analysis of how acute alcohol consumption and injury or collision risk increase together. *Drug and Alcohol Dependence*, 110(1–2), 108–116.

Taylor, B., & Rehm, J. (2012). The relationship between alcohol consumption and fatal accidents. *Alcohol Clinical & Experimental Research*, 36(10), 1827–1834.

Tchir, J. (2018). Is there a safe number of drinks I can consume before driving? *Globe and Mail*, June 4, 2018.

Terhune, K. W., & Fell, J. C. (1982). The Role of Alcohol Marijuana, and Other Drugs in the Accidents of Injured Drivers. In *25th Annual Conference of the American Association for Automotive Medicine* (pp. 15). San Francisco: US Department of Transportation.

Terhune, K. W., Ippolito, C. C., Hendricks, D. L., Michalovic, S. C., Bogema, S. C., Santiga, P., ... Preusser, D. F. (1992). *The Incidence and Role of Drugs in Fatally Injured Drivers*. Washington, DC.

Thomas, C. D., Cameron, A., Green, R. E., Bakkenes, M., Beaumont, L. J., Collingham, Y. C., & Hughes, L. (2004). Extinction risk from climate change. *Nature, 427* (6970), 145-148.

Todorovic, M.S., Todorovic, D.V., Matejic, S., Minic, Z.S., Stankovic, V., Preljevic, I., Hajrovic, S., & Savic, S.N. (2012). The role of alcohol in fatal work related injuries. *HealthMED*, 6, 649-653.

Toennes, S. W., Steinmeyer, S., Maurer, H. J., Moeller, M. R., & Kauert, G. F. (2005). Screening for drugs of abuse in oral fluid - Correlation of analysis results with serum in forensic cases. *Journal of Analytical Toxicology*, 29(1), 22–27.

Toennes, S.W., Ramaekers, J.G., Theunissen, E.L., Moeller, M.R., & Kauert, G.F. (2010). Pharmacokinetic properties of delta9-tetrahydrocannabinol in oral fluid of occasional and chronic users. *Journal of Analytical Toxicology*, 34(4), 216–21.

Tyrrell, I. (1997). The US prohibition experiment: Myths, history and implications. *Addiction, 92*(11), 1405–1409.

U.S. Department of Transportation. (2004). *Drugs and Human Performance Fact sheets* (Report No. DOT HS 809 725). National Highway Traffic Safety Administration.

U.S. Department of Transportation. (2015). Critical reasons for crashes investigated in the National Motor Vehicle Causation Survey: Traffic Safety Facts.

Unger, N., Bond, T. C., Wang, J. S., Koch, D. M., Menon, S., Shindell, D. T., & Bauer, S. (2010). Attribution of climate forcing to economic sectors. *Proceedings of the National Academy of Sciences*, 107(8), 3382-3387.

Van Laar, M. W., van Willigenburg, A. P. P., & Volkerts, E. R. (1993) Over-the-road and simulated driving: Comparison of measures of the hangover effects of two benzodiazepine hypnotics. In Utzelmann H.D., Erghaus G., Kroj G., Verlag T.Ü.V., Rheinland G., and Köhn B.H. (Eds.), *Alcohol, Drugs, and Traffic Safety* (pp. 672-677).

Vanlaar, W., Robertson, R., Marcoux, K., Mayhew, D., Brown, S., & Boase, P. (2012). Trends in alcohol-impaired driving in Canada. *Accident Analysis and Prevention*, *48*, 297–302.

Verster, J. C., et al (2009). The alcohol hangover research group consensus statement on best practice in alcohol hangover research. *Current Drug Abuse Reviews, 119*(11), 2658–2666.

Verster, J. C., Bervoets, A. C., de Klerk, S., Vreman, R. A, Olivier, B., Roth, T., & Brookhuis, K. A. (2014). Effects of alcohol hangover on simulated highway driving performance. *Psychopharmacology, 231*(15), 2999–3008.

Verstraete, A., Knoche, A., Jantos, R., Skopp, G., Gjerde, H., Vindenes, V., & Lillsunde, P. (2011). *Per se limits: methods of defining cut-off values for zero tolerance.* DRUID. Ghent: Ghent University.

Verstraete, A.G. (2004). Detection times of drugs of abuse in blood, urine, and oral fluid. *Therapeutic Drug Monitoring, 26,* 200-206.

Vindenes, V., Lund, H.M.E., Andresen, W., Gjerde, H., Ikdahl, S.E., Christophersen, A.S., & Øiestad, E.L. (2012). Detection of drugs of abuse in simultaneously collected oral fluid, urine and blood from Norwegian drug drivers. *Forensic Science International*, 219(1-3), 165-171.

Vingilis, E., & Macdonald, S. (2002). Review: Drugs and Traffic Collisions. *Traffic Injury Prevention* (3).

Volkerts, E. R., van Lair, M., W. (1993) A methodological comparative study of over- the-road and simulated driving performance after nocturnal treatment with lormetazepam 1 mg and oxazepam 50 mg. In Utzelmann H.D., Berghaus G., Kroj G., Verlag T.Ü.V., Rheinland G., and Köhn B.H. (Eds.), *Alcohol, Drugs, and Traffic Safety* (pp. 664-671.).

Volkow, N. D., Baler, R. D., Compton, W. M., & Weiss, S. R. (2014). Adverse health effects of marijuana use. *The New England Journal of Medicine, 5*(370), 2219–27.

Wagenaar, A.C., & Maldonado-Molina M.M. (2007) Effects of drivers' license suspension policies on alcohol-related crash involvement: long-term follow-up in forty-six states. Alcoholism: *Clinical & Experimental Research, 31*, 1399-406.

Wald, N. J., & Bestwick, J. P. (2014). Is the area under an ROC curve a valid measure of the performance of a screening or diagnostic test? *Journal of Medical Screening,* 21(1), 51–56.

Walsh, G., & Mann, R. (1999). On the high-road: Driving under the influence of cannabis in Ontario. *Canadian Journal of Public Health*, *90*(4), 260–263.

Walsh, J.M., Verstraete, A.G., Huestis, M.A., & Mørland, J. (2008). Guidelines for research on drugged driving. *Addiction, 103*(8), 1–16.

Watson, S. J., Benson, J., & Joy, J. E. (2013). Marijuana and Medicine: Assessing the Science Base. *Marijuana and Medicine*, 57(June 2000), 170.

Watson, T. M., & Mann, R. E. (2016). International approaches to driving under the influence of cannabis: A review of evidence on impact. *Drug and Alcohol Dependence, 169*, 148–155.

Wettlaufer, A., Florica, R. O., Asbridge, M., Beirness, D., Brubacher, J., Callaghan, R., ... Rehm, J. (2017). Estimating the harms and costs of cannabis-attributable collisions in the Canadian provinces. *Drug and Alcohol Dependence, 173*, 185–190.

White, M. (2017) Cannabis and Road Crashes: A close look at the best epidemiological evidence. November 1, Final Report.

Wiese, J., & Shlipak, M. (2001). The alcohol hangover. *Annals of Internal Medicine, 134*(6), 534.

Wigmore, J. (2014). The forensic toxicology of alcohol and best practices for alcohol testing in the workplace. *Canadian Nuclear Safety Commission*.

Williamson, A. M., Feyer, A. M., Mattick, R. P., Friswell, R., & Finlay-Brown, S. (2001). Developing measures of fatigue using an alcohol comparison to validate the effects of fatigue on performance. *Accident Analysis and Prevention*, 33(3), 313–326.

Wilson, F. A., Stimpson, J. P., & Pagán, J. A. (2014). Fatal crashes from drivers testing positive for drugs in the U.S., 1993-2010. *Public Health Reports, 129*(4), 342–

Wolff, K., & Johnston, A. (2014). Cannabis use: a perspective in relation to the proposed UK drug-driving legislation. *Drug Testing and Analysis*, 6(1–2), 143–54.

Wolff, K., Farrell, M., Marsden, J., Monteiro, MG., Ali, R., Welch, S., & Strang, J. (1999). A review of biological indicators of illicit drug use, practical considerations and clinical usefulness. *Addiction*, 94(9), 1279–98.

Wong, K., Brady, J. E., & Li, G. (2014). Establishing legal limits for driving under the influence of marijuana. *Injury Epidemiology, 1*(26), 1-8.

Wood, E. (2016). Why a 5 ng / mL THC limit is bad public policy - and the case for Tandem per se DUID legislation. *The Journal of Global Drug Policy and Practice, 10*(1–21).

Woodall, K. L., Chow, B. L. C., Lauwers, A., & Cass, D. (2015). Toxicological findings in fatal motor vehicle collisions in Ontario, Canada: a one-year study. *Journal of Forensic Sciences, 60*(3), 669–674.

World Health Organization. (2009). *Polydrug use: patterns and responses. Luxembourg: Publications Office of European Union.* Retrieved from http://www.emcdda.europa.eu/publications/selectedissues/polydrug-use

World Health Organization. (2014). *Global status report on noncommunicable diseases, 2014.* Retrieved from http://apps.who.int/iris/bitstream/10665/148114/1/9789241564854_eng.pdf?ua=1

World Health Organization. (2015). *Global status report on road safety 2015.* Retrieved from http://www.who.int/violence_injury_prevention/road_safety_status/2015/en/

World Health Organization. (2017). *management of substance abuse: facts and figures* http://www.who.int/substance_abuse/facts/global_burden/en/

World Health Organization. (undated). Management of substance abuse. Retrieved from http://www.who.int/substance_abuse/facts/cannabis/en/

Wright, V. (2010). *Deterrence in criminal justice: Evaluating certainty vs. severity of punishment.* Sentencing Project: Research and advocacy for reform.

Wutke, S. (2014). *History of the Breathalyzer.* Guardian Interlock. Retrieved from https://guardianinterlock.com/blog/history-breathalyzer/)

Yesavage, J.A., Leirer, V.O., Denari, M., & Hollister, L.E. (1985). Carry-over effects of marijuana intoxication on aircraft pilot performance: a preliminary report. *American Journal of Psychiatry, 142*(11), 1325–1329.

Zador, P. L., Krawchuk, S. A., & Voas, R. B. (2000). Relative Risk of Fatal Crash Involvement by BAC, Age and Gender. Washington.

Zakhari, S. (2006). Overview: How is alcohol metabolized by the body? *Alcohol Research & Health, 29*(4), 245-254.

Zeisser, C., Thompson, K., Stockwell, T., Duff, C., Chow, C., Vallance, K., & Lucas, P. (2012). A 'standard joint'? The role of quantity in predicting cannabis-related problems. *Addiction Research & Theory, 20*(1), 82-92.

Zuba, D. (2008). Accuracy and reliability of breath alcohol testing by handheld electrochemical analysers. *Forensic Science International, 178*(2–3), 29-33.

Glossary

Accuracy— Accuracy refers to the ability of a diagnostic test to correctly classify people as impaired or not impaired, assuming a given definition of impairment. It is calculated by the proportion of true positives and negatives divided by all the cases being examined.

Different indicators of validity of a biological test for impairment

	Impaired	**Not impaired**
Test: Positive	True positive	False positive
Test: Negative	False negative	True negative

Alcohol ignition interlock—An ignition interlock system is a Breathalyzer installed in a vehicle where a driver must provide a negative breath sample (under a prescribed threshold) in order to start the automobile.

Back-extrapolation—Extrapolation is an estimation of a value based on extending a known sequence of values beyond the area that is observed. Back extrapolations involve estimating values at a prior point in time.

Bias—A systematic error in the design or conduct of a study that leads to an erroneous association between the exposure and an outcome, such as crashes.

Blinded study/double-blinded study—A study in which subjects do not know what treatment they are receiving (participant blinding) or where the experimenters do not know what treatment the subjects are receiving (experimenter blinding). In a **double-blind** study, neither the participants nor the experimenters know who is receiving a particular treatment.

Blood Alcohol Content (BAC)—The percentage of alcohol in the bloodstream.

Cannabidiol (CBD)—A non-psychoactive compound from cannabis which is reported to have medicinal properties.

Carboxy-THC (THCCOOH)—A metabolite in THC that is typically detected with urinalysis and used to identify drug users rather than those impaired. Like THC, carboxy-THC can be stored in fat cells for days, and possibly weeks in heavy daily users.

Case-control study—A study that compares subjects (i.e. cases) who have a disease or condition (i.e.. traffic crashes are considered a condition) to those without a condition (controls), in terms of exposure to a risk factor (e.g. alcohol or cannabis). The purpose is to determine the relationship between the risk factor and the condition through calculations of odds ratios.

Causation (general causation)—Causality means that a change in one variable will alter another variable. General causation can be determined for groups of people using epidemiological methods. There are three requirements in order to demonstrate general causation in epidemiological studies: the suspected cause must precede the suspected effect in time, a statistical relationship between two variables of interest must exist, and the observed empirical relationship cannot be explained by a third variable.

Central tendency—A statistical measure that describes the way in which a group of data clusters around a central value. There are three measures of central tendency: the mean, the median and the mode.

Chromatography and spectrometry—Gas chromatography mass spectrometry (GC/MS) is an instrumental technique, comprising of a gas chromatograph (GC) coupled to a mass spectrometer (MS), by which compounds can be quantified. A combination of these methods is considered the gold standard and referred to as confirmatory tests.

Coefficient of determination (r^2)—A squared correlation coefficient (r) which can be interpreted as the percent of variation in one variable that is shared with a second variable.

Coefficient of variation (CV)—This is the ratio of the standard deviation to the mean of a sample. The CV is useful to compare the dispersion (variability) among variables that have different units of analysis and means. Variability is standardized whereby sample distributions with different units of analysis can be compared directly in terms of their variability. Low CVs indicate low variability.

Cohort study—A group of people followed-up over time assess the risk of developing a particular health outcome.

Compound—A substance composed of atoms from more than one element, held together by chemical bonds.

Confidence interval (CI)—A range that means that if the sample was drawn with exactly the same methods, the sample characteristics would lie within this range about 19 out of 20 times.

Confounding variable—A confounding variable is causally related to a condition or serves as a proxy measure for unknown causes and is associated with the exposure under study but is not a consequence of exposure (see Kelsey et al., 1996).

Continuous variable —A variable that can have many values.

Controlled dosing studies—Studies in which subjects are administered substances, and concentration levels with drug tests are measured at different time intervals after use.

Conversion factors—A conversion factor is defined as a ratio of two numbers that make them equivalent.

Correlation—The extent to which two variables are statistically related to one another.

Correlation coefficient (r)—A statistical measure of the degree to which changes of the values of one variable are related to changes of the values of another variable. The value of r can range from -1 to +1, with the sign indicating the direction of the relationship. A correlation of r=1 is perfect and 0 is no relationship (see also *Coefficient of determination*).

Correlation of determination (r^2) – The percentage of variation in one variable that can be explained by another variable.

Counterbalanced design—A research design in which subjects receive two or more treatment conditions balanced in opposite order. The design often is used to reduce the likelihood of learning effects and reduces bias.

Cross-over trial—A cross-over study (trial), is an experimental randomized study in which subjects receive a sequence of different treatments (or exposures) in a sequence that is counterbalanced.

Culpability study—Culpability (responsibility) design is a type of observational study in which all drivers have been involved in a crash (usually fatal), and drug tests are taken to assess the presence of drugs. Odds ratios are then calculated to assess the association between being responsible for a crash and testing positive.

Deterrence theory—This theory posits that the certainty, swiftness and severity of punishments for offenders will change unwanted behaviours (Ross, 1982). The cornerstone of deterrence is punishment. According to deterrence theory, if undesirable acts are punished then people will be less likely to engage in those behaviours in the future (i.e. specific deterrence) and the larger population will be less likely to engage in behaviours based on fear of being punished (i.e. general deterrence).

Dispersion—Dispersion (**variability**) is the extent to which a variable is varied. It is statistically measured by standard deviation, variance and coefficient of variation.

Double-blind experimental approach—See *Blinded study*.

Drug Recognition Experts (DRE)—Also called the *Drug Evaluation and Classification Program* (DEC). Developed in the late 1970s by the Los Angeles Police Department, this is a standardized procedure that is used to identify both individuals under the influence of drugs, and the type of drug causing the observable impairment (Beirness et al., 2007).

Effect size—This refers to the actual strength of statistical relationships, whether they are correlations, OR or RR. Statistical significance (i.e. $p<.05$) is best thought of as an essential test requirement, and if significant, examines the effect size to assess whether the relationship is meaningful (i.e. has practical implications). In this book, I have defined effect sizes for that I consider weak, moderate or strong different levels of measurement. More details about effect sizes can be found in Chapter 7, Epidemiological guidelines for valid findings, (4) Strength of Association.

Ethanol, also called absolute alcohol—The intoxicating agent found in alcoholic beverages, such as beer, wine and spirits. The amount of ethanol varies considerably among these beverages with average contents of 5% for beers, 12% for wines and 40% for spirits.

Experiments—An operation or procedure carried out under controlled conditions.

Equilibrium—An equal balance.

External validity—The degree to which the findings of a study can be generalized or translated to groups that did not participate in the study. If the results of the study apply to the "real-world" observational studies (field studies), they tend to have better external validity than laboratory studies.

False negative—A test result which incorrectly indicates that a particular condition is absent.

False positive—A test result which incorrectly indicates that a particular condition is present.

GC/MS—Gas chromatography and mass spectrometry, a drug testing analytical approach, sometimes referred to as the gold standard.

GC/MS/MS—Gas chromatography coupled with tandem mass spectrometry.

Gold standard for drug tests—These are confirmatory drug tests using chromatography and spectrometry methods to detect a specific compound associated with prior exposure to drugs.

Gold standard for impairment—An external source of truth regarding the magnitude of deficits known to be associated with increased crash risk.

Harm reduction—A set of practical strategies which are aimed at reducing the negative consequences of drug use without requiring abstinence as a goal. An analogy of this term has been applied to approaches that aim at reducing the likelihood of crashes and injuries to drivers, accepting the fact that people drive, and that driving can be dangerous.

Heterogeneity—Variability in the intervention effects of different studies.

Hydroxy-THC (11-OH-THC)—Another major metabolite of THC. Compounds and metabolite concentrations are typically expressed in nano-grams per millilitre (ng/mL). The focus of this book will be mainly on detection of THC and these metabolites.

Immunoassay—A screening test that identifies compounds through an antibody test, a test that uses the binding of antibodies to antigens to identify and measure certain substances. It is substantially less valid than confirmatory analyses as it cannot accurately distinguish compounds with similar molecular structures.

Impairment—Refers to a specific threshold at which driving a vehicle due to performance deficits is dangerous for most people. For alcohol, this threshold is .05%.

Imputed—The replacement of missing values in a data set based on known characteristics of variables from valid cases.

Independent measures—Each observation on a variable should be independent. This means that multiple observations of the same subject would violate the assumption of independence and should not be used for statistical procedures, such as correlations. One observation should not influence or affect the value of other observations, and multiple observations of the same individuals (i.e. within persons) will tend to be more similar than those of different individuals (i.e. between persons). Violations of this assumption will typically erroneously increase an effect size and sample size, thereby increasing the likelihood of statistical significance.

Indictable offence—A more serious offence under the Criminal Code of Canada.

Internal validity—The degree to which a study is free from bias to allow for strong inferences between suspected cause and effects. Randomized double-blind studies are the gold standard in terms of internal validity. Threats to internal validity represent various explanations of how the results of a study might not be internally valid, based on the designs. These include potential flaws, such as selection (differences found are due to fundamental differences in the group), history (competing interventions caused the change), experimenter bias, testing effects (both practice effects and Hawthorne effect) and statistical regression.

Kappa—Statistics often used to assess the degree of agreement between a test and a gold standard, both with dichotomous outcomes. Gordis (2014) has provided an interpretation of Kappa values as follows: > .75 excellent agreement, .40 to .75 intermediate to good agreement, < .40 poor agreement.

Laboratory studies—Research studies in controlled environments, which may or may not have randomized conditions.

LC/MS/MS—Liquid chromatography and tandem mass spectrometry, confirmatory methods typically used for detecting specific compounds from oral fluids.

Mens rea—Criminal intent for a particular behaviour that is a crime.

Metabolite—A compound that is formed by the body's metabolism of a drug.

Multivariate analysis—The statistical assessment of the relationship between two or more variables on a dependent variable. Typically, a multivariate analysis would include a main independent variable, thought to be causally related to an outcome, and one or more confounding variables. The primary purpose of this analysis is to adjust (i.e. remove) the effect of confounding variables in explaining a particular outcome.

Non-significant—Findings from a statistical test where we must conclude that there is no relationship between two or more variables.

Odds ratio (OR)—The ratio of the odds of getting a disease/health outcome (e.g. crash) if exposed (e.g. cannabis positive) to the odds of getting a disease if unexposed. It is calculated by the formula a*d/c*b (see table below).

Cells used for the calculation of Odds Ratios

Outcome	Exposed	Unexposed
Disease (e.g. crash, responsible)	a	b
No disease (e.g. no crash, not responsible)	c	d

Observational studies—Studies of human behaviour in the real world.

One-tailed statistical test—A statistical test that only examines one direction of a relationship.

Operational definition—A clear explanation of how to measure or detect something.

Per se law—Per se means "by itself." A per se law for a BAC at or above .08% alcohol means that a person found at this level is considered intoxicated. No further evidence is necessary to demonstrate intoxication.

Performance deficits—Defined more broadly to include any decrements in psychomotor, perceptual or cognitive functioning. Performance deficits do not necessarily imply impairment.

Placebo—A substance of which subjects are not aware that it is inactive.

Plasma—The liquid portion of blood that remains after red blood cells, white blood cells, platelets and other cellular components are removed.

Point-of-collection tests—A fast drug test that gives an indication within 10 minutes of drug exposure.

Positive predictive value—The probability that a person who tests positive truly has a particular condition, such as being impaired.

Prevalence—The proportion of a particular population with a particular condition.

Probability values—The likelihood that the results of a statistical test could be due to chance.

r and r^2—See Correlation coeffcient (r) and Coefficient of determination (r^2).

Random assignment—A study in which each subject has an equal chance of being assigned to different treatment conditions.

Random breath testing—Stopping drivers and demanding a breath test without reasonable suspicion that the drivers have been drinking.

Receiver Operating Characteristics (ROC)—An analytic approach that compares sensitivity and specificity of a test so that the overall accuracy of the test is maximized.

Recidivism—The likelihood that someone convicted of a crime will re-offend.

Relative risk function or model—A combination of numerous relative risk point estimates to illustrate a dose-response relationship.

Relative risk or **risk ratio** (**RR**)—The ratio of the probability of an event (e.g. crash) in an exposed group (e.g. cannabis positive) to the probability of the event occurring in a non-exposed group (e.g. cannabis negative). It is calculated by the formula: a/(a+b)/c/(c+d) (see table below). Although the formulas for relative risk and odds ratio are different, the terms are sometimes used synonymously. Technically, relative risks should be calculated from cohort studies.

Cells used for the calculation of relative risk

Outcome	Exposed	Unexposed	Total
Disease (e.g. crash)	a	b	a+b
No disease (e.g. no crash)	c	d	c+d

Reliability—The degree to which an assessment tool produces consistent findings.

Response bias—Bias created in a study in which the respondents are not representative of all the subjects who could have participated.

Selection bias—This type of bias is created, typically due to non-respondents, when drug users in the control group are less likely to participate in a study than the cases. It is particularly an issue where consent to participate in the study is required from the control subjects but not the cases. This limitation will also have the effect of creating inflated odds ratios (higher than reality) because drug users will be under-represented in the control group as they will be less likely to participate.

Selective breath testing—In order for police to demand a driver takes a Breathalyzer test, reasonable grounds that a driver has been drinking are needed.

Sensitivity—Defined as the ability of the test to identify correctly those who have a specific disease (Gordis, 2009; p. 86). In this book, sensitivity is usually referred to as the ability of a test to correctly identify those who are impaired.

Sequential tests—Two tests are used to diagnose a condition where only those who are positive on the first test are subject to a second test that assesses the condition using a different approach. Sequential tests are more appropriate as a diagnostic than screening tests, and increase specificity (i.e. reduce false positives) compared to a single test.

Serum—Serum is the part of blood which is similar in composition with plasma but excludes clotting factors of blood.

Specificity—Ability of a test to identify correctly those who do not have the disease (Gordis, 2009; p. 86). In this book, specificity is usually referred to as the ability of a test to correctly identify those who are **not** impaired.

Standard deviation—This is a measure used to define the extent of variation or dispersion of a sample. In a normal distribution, about 68% of the observations lie above and below the mean (i.e. average).

Standardized Field Sobriety Tests (SFST)—Three tests used by police to assess behavioural symptoms of alcohol impairment: the one-legged stand, walk and turn and alcohol gaze nystagmus. These tests have been validated against a BAC cut-off of .08%.

Statistical significance—The likelihood that a result or relationship is caused by something other than random chance. The probability value is the likelihood that a given strength of a relationship is due to chance, and for this value to be significant, it should be below 5% or $p<.05$. The likelihood of probability values being significant is based on both the sample size and the variability in the measures. This means that very weak relationships between variables can become significant with a large enough sample.

Summary offence—These are less serious Criminal Code of Canada offences.

Temperance movements—A social movement against the consumption of alcoholic beverages.

Tetrahydrocannabinol (THC)—The chemical compound responsible for most of cannabis's psychoactive effects.

Titration—Adjusting a dosage, typically by smoke inhalation, to obtain the optimal dose of cannabis, which varies considerably from person to person, depending on the level of high or medicinal effect desired.

Two-tailed statistical test—A statistical test that examines two directions of a relationship.

Validity—The degree to which a test measures what it was designed to measure. Validity of the diagnostic test is defined as the ability to distinguish between those who have a disease and those who have not (Gordis, 2009).

Variability—Variability is the extent to which data points in a variable are spread apart.

Whole blood—Blood drawn directly from the body from which no component, such as white and red cells, plasma or platelets, has been removed.